The rise in the concentration of CO_2 in the atmosphere since the start of industrialisation, and the global warming associated with this greenhouse gas, has stimulated research into the response of plants to elevated levels of CO_2. Much of this work has been carried out in controlled environments in simple experiments of short duration which provide limited information about long-term effects on vegetation. In contrast, CO_2-emitting mineral springs provide a unique opportunity to consider vegetation which has endured over many generations at naturally elevated levels of CO_2. This volume presents findings from a range of sites, confirming the potential of these natural laboratories in the investigation of this important aspect of climate change.

Plant responses to elevated CO_2

Plant responses to elevated CO$_2$

Evidence from natural springs

EDITED BY

A. Raschi, F. Miglietta,

R. Tognetti and P. R. van Gardingen

CAMBRIDGE
UNIVERSITY PRESS

CAMBRIDGE UNIVERSITY PRESS
Cambridge, New York, Melbourne, Madrid, Cape Town, Singapore, São Paulo, Delhi

Cambridge University Press
The Edinburgh Building, Cambridge CB2 8RU, UK

Published in the United States of America by Cambridge University Press, New York

www.cambridge.org
Information on this title: www.cambridge.org/9780521582032

First published 1997
This digitally printed version 2008

A catalogue record for this publication is available from the British Library

Library of Congress Cataloguing in Publication data
Plant responses to elevated CO_2 evidence from natural springs / edited by
A. Raschi . . . [*et al.*].
    p. cm.
  Includes index.
  ISBN 0 521 58203 2 (hb)
  1. Plants, Effect of carbon dioxide on. 2. Carbon dioxide–Environmental aspects.
3. Springs. I. Raschi, A.
QK753.C3P58   1997
581.7–dc21   97-8030 CIP

ISBN 978-0-521-58203-2 hardback
ISBN 978-0-521-08729-2 paperback

Contents

Contents

Contributors

A D'Annibale, Dipartimento di Agrobiologia e Agrochimica, Universita'di Viterbo,
   Via S C De Lellis, I-01100 Viterbo, Italy

M Badiani, Dipartimento di Agrobiologia e Agrochimica, Universita'di Viterbo,
   Via S C De Lellis, I-01100 Viterbo, Italy

J D Barnes, Department of Agricultural and Environmental
Science, Ridley Building, University of Newcastle-upon-Tyne, Newcastle-upon-Tyne,
   NE1 7RU, UK

JD Beerling, Department of Animal and Plant Sciences, University of Sheffield,
   S10 2TN UK

I Bettarini, IATA-CNR, P le delle Cascine 18, I- 50144 Firenze, Italy

SH Byari, Faculty of Meteorology, Environment and Arid-land Agriculture,
   King Abdulazziz University, Jeddah, Saudi Arabia

AC Cook, Department of Biology & Global Change Research Group,
   San Diego State University, San Diego, California 92182, USA

MF Cotrufo, Division of Biological Sciences, Institute of Environmental and Biological
   Sciences, Lancaster University, Lancaster LA1 4YQ, UK

JR Ehleringer, Department of Biology, University of Utah, Salt Lake City,
   Utah 84112, USA

ZH Enoch, Agricultural Research Organisation, The Volcani Center, Institute of Soils
   and Water, Bet Dagan, Israel

S Esposito, Dipartimento di Biologia Vegetale, Università di Napoli Federico II,
   Via Foria 223, 80139 Napoli, Italy

G Etiope, Istituto Nazionale di Geofisica, Via di Vigna Murata, 605-00143 Roma, Italy

MC Fordham, Department of Agricultural and Environmental Science, Ridley Building,
   University of Newcastle-upon-Tyne, Newcastle-upon-Tyne, NE1 7RU, UK

J Grace, Institute of Ecology & Resource Management, The Darwin Building, Mayfield
   Road, Edinburgh EH9 3JU, UK

HG Griffiths, Department of Agricultural and Environmental Science, Ridley Building,
   University of Newcastle-upon-Tyne, Newcastle-upon-Tyne, NE1 7RU, UK

P Grossoni, Laboratorio Botanica Forestale, Dipartimento Biologia Vegetale, Università di
   Firenze, P le delle Cascine 28, I-50144, Firenze, Italy

P Ineson, Merlewood Research Station, Grange-over-Sands, Cumbria LA11 6JU, UK

CE Jeffree, Science Faculty Electron Microscopy Unit, University of Edinburgh, UK

JD Johnson, School of Forest Resources and Conservation, University of Florida, 326 Newins-Ziegler Hall, Gainesville, Florida 32611, USA

S Lombardi, Dipartimento di Scienze della Terra, Università "La Sapienza", P le A Moro, 5-00185 Roma, Italy

F Manes, Dipartimento Biological Vegetale, Università "La Sapienza", P le A Moro, 5-00185 Roma, Italy

M Martini, Department of Earth Sciences, Università di Firenze, ItalyV Di Martino-Rigano, Dipartimento di Biologia Vegetale, Università di Napoli Federico II, Via Foria 223, 80139 Napoli, Italy

C Di Martino, Dipartimento di Biologia Vegetale, Università di Napoli Federico II, Via Foria 223, 80139 Napoli, Italy

M Michelozzi, Istituto Miglioramento Genetico delle Piante Forestali, Consiglio Nazionale delle Richerche, via Atto Vannucci 13, 50134 Firenze, Italy

F Miglietta, IATA-CNR, P le delle Cascine 18, I- 50144 Firenze, Italy

M Mousseau, Laboratoire d'Ecologie Végétale, URA 1492, Université Paris-sud, 91405 Orsay Cedex, France

WC Oechel, Department of Biology & Global Change Research Group, San Diego State University, San Diego, California 92182, USA

AR Paolacci, Dipartimento di Agrobiologia e Agrochimica Universita'di Viterbo, Via S C De Lellis, I-01100 Viterbo, Italy

E Paoletti, CSPSLM-CNR, P le delle Cascine 28, I- 50144 Firenze, Italy

SL Philips, Department of Biology, University of Utah, Salt Lake City, Utah 84112, USA

J Piñol, CREAF, Universitat Autònoma de Barcelona, 08193 Bellaterra, Spain

A Raschi, IATA-CNR, P le delle Cascine 18, I- 50144 Firenze, Italy

C Rigano, Dipartimento di Biologia Vegetale, Università di Napoli Federico II, Via Foria 223, 80139 Napoli, Italy

JC Sabroux, Institut de Protection et de Sûreté Nucléaire/DPEI/SERAC, Centre d'Etudes de Saclay, 91191 Gif sur Yvette, Cedex, France and UATI, 1 rue Miollis, 75732 Paris Cedex 15, France

DR Sandquist, Department of Biology, University of Utah, Salt Lake City, Utah 84112, USA

F Selvi, Dipartimento di Biologia Vegetale, Università di Firenze, Via la Pira 4, I-50121, Firenze, Italy

B Sveinbjornsson, Institute of Biology, University of Iceland, 108 Reykjavik, Iceland and University of Alaska, Anchorage, Alaska 99508, USA

J Terradas, CREAF, Universitat Autónoma de Barcelona, 08193 Bellaterra, Spain

R Tognetti, Istituto Miglioramento Genetico delle Piante Forestali, Consiglio Nazionale delle Richerche, via Atto Vannucci 13, 50134 Firenze, Italy

P R van Gardingen, Institute of Ecology & Resource Management, The Darwin Building, Mayfield Road, Edinburgh EH9 3JU, UK

V Vona, Dipartimento di Biologia Vegetale, Università di Napoli Federico II, Via Foria 223, 80139 Napoli, Italy

F I Woodward, Department of Animal and Plant Sciences, University of Sheffield, S10 2TN, UK

Preface

There has been substantial discussion by politicians, scientists and the general public about the likely effects of the changing composition of the atmosphere of the Earth. The CO2 concentration of the Earth's atmosphere continues to increase in response to land-use change and the burning of fossil fuels. There is no question that this increase is directly linked to the ever increasing global human population and our demand for energy. This is one of the most easily quantified aspects of global environmental change.

It is well accepted that the enhanced CO_2 concentrations associated with global change will directly influence the process of photosynthesis in most species of plants. Early observations suggested that increased rates of photosynthesis would lead to increased growth rates of plants. There was a proliferation of experimental studies investigating the effects of CO_2 on plants. As more data were collected these became increasingly difficult to interpret. The majority of studies were conducted under artificial conditions in growth rooms, glasshouses or open-top chambers. The great deficiency of most studies was that vegetation was exposed for relatively short periods of time. There seemed to be no way forward that would permit the study of the long-term effects of exposure of vegetation to enhanced CO_2 concentrations.

In the early 1990s a few groups of scientists faced with this problem suggested using sites with naturally enhanced atmospheric levels of CO_2 as natural experiments. There followed a number of scientific programmes in several locations around the world where vegetation was exposed to enhanced CO2. Most of the work has since been conducted at sites with natural mineral springs, mainly for the reason that these locations have relatively low levels of atmospheric pollutants associated with the CO_2 emissions. This volume describes much of this work.

The publication resulted from an international workshop on the topic of Carbon Dioxide Springs and their Use in Biological Research, held at the small village of San Miniato in Tuscany, central Italy. A varied group of international scientists convened from 14–17 October 1993. These chapters represent their findings and conclusions. The meeting was hosted by the Environmental Physiology Groups of the British Ecological Society and the Society for Experimental Biology in conjunction with the Global Change in Terrestrial Environment group (GCTE) of the International Geosphere Biosphere Programme (IGBP). Locally the meeting was sponsored by CNR-IATA and

CeSIA, Accademia dei Georgofili, the Comune di Rapolano Terme and the local bank, Cassa di Risparmio di S. Miniato.

The production of this volume has involved a team of people who have been of invaluable assistance to the editors. Firstly we thank all of the authors for their contributions and generally trying where possible to keep to deadlines. The workshop was run with the assistance of A. Falchi. The team at Cambridge University Press have maintained their good humour throughout the production of the volume; the guidance from Dr Maria Murphy and Dr Alan Crowden at CUP has been especially helpful. To everyone involved with this project, we express our collective appreciation.

February 1997
A. Raschi
F. Miglietta
R. Tognetti
P.R. van Gardingen

Sites of naturally elevated carbon dioxide

J. GRACE AND P.R. VAN GARDINGEN

Institute of Ecology and Resource Management, The Darwin Building, Mayfield Road, Edinburgh, EH9 3JU, UK.

SUMMARY

The continuing rise of CO_2 concentration in the atmosphere since the start of industrialisation, and the associated global warming, pose interesting questions about the response of plants to CO_2. However, in the course of evolution, plants have been exposed to a much wider range of CO_2 concentrations than we have seen in the last 200 years. It is likely that when photosynthesis evolved, around 3.8 billion years ago, the earth's atmosphere was CO_2 rich, just as Mars and Venus are today. In order to understand the adaptability of plants to changing CO_2, there are several possible experimental approaches. The presence of CO_2 springs in several parts of the world provides a natural laboratory for such studies, with the important advantage that plants growing at such sites will have been exposed to elevated CO_2 for many generations.

The CO_2 concentration of the earth's atmosphere when photosynthesis began, some 3.8×10^9 years ago, is presumed to have been in the range 90-98 %, similar to that found currently on the lifeless sister planets Mars and Venus (Emiliani, 1992; Raven, 1995). This high CO_2 atmosphere originated from the outgassing of the planet's crust, which continues today albeit at a diminished rate in volcanic regions of the world.

Photosynthesis in those early days of the earth's history is believed to have been achieved by prokaryote filamentous cyanobacteria, forming crusts as seen now at coastal areas in warm parts of the world. The rates of photosynthesis achieved by these organisms may have been quite high, as a result of the large diffusion gradient from the atmosphere to the sites of carboxylation. There is no reason to believe that the biochemistry of ancient photosynthesis was any different from what we find today in C_3 plants. Indeed, ribulose bisphosphate carboxlyase (rubisco) has several features that suggest great evolutionary age (Raven, 1995). These include its excessive size in relation to its task, with a good deal of redundancy suggestive of vestigal functional components, its associated oxygenase activity which competes with its carboxylating activity and makes the enzyme seem unsuitable for the current environment of high O_2 and low CO_2, and its widespread distribution among taxa.

Rubisco has survived great changes in atmospheric composition, as the CO_2 concentration has been drawn down to a few hundred parts per million as a result of a combination of processes aided by photosynthesis, including deposition to the ocean, sedimentation to the ocean floor with subsequent formation of carbonate rocks, and the deposition of carbon in soils and fossil fuels. Rather little is known about the adaptability of rubisco to changes in ambient CO_2 even though it is the subject of much research. Naturally enriched CO_2 sites, especially those at mineral springs, offer an exciting opportunity to examine the contrasts between species grown for extended periods under a wide range of concentrations.

The possibility that today's vegetation has been growing faster over the last century or so as the result of CO_2 fertilisation from anthropogenic emissions, and the real likelihood that forests and grasslands are net accumulators of carbon on a global scale (Körner, 1993; Grace et al., 1995) has prompted a close analysis of the response of plants to a CO_2-enriched atmosphere. The question of the acclimation (short-term) and adaptation (long-term) to changes in CO_2 is important in any attempts to predict how plants and ecosystems are likely to respond to the current trend of increasing atmospheric CO_2 concentrations. Recent models of the response of ecosystems to global change have usually presumed that the CO_2 response is adequately represented by a well-known model of photosynthesis (Farquhar, von Caemmerer & Berry, 1980; Lloyd et al., 1995; Wang & Polglase 1995), and acclimation and adaptation have been ignored. In reality, there are several possible negative feedbacks which come into play, which must be superimposed on this pattern. These include the frequently reported (but not invariable) reduction in stomatal conductance at high CO_2 concentration, responses at the biochemical level, the whole plant level, and also at the ecosystem level. Currently, it is unclear which of these is the most important.

Postulated responses at the biochemical level include the downregulation of rubisco activity and the regenerative capacity of ribulose bisphosphate (Bowes, 1993). Inferences about such downregulation can be made, to some extent, from the determination of A/C_i curves. At the whole plant level, the carbon allocation patterns are liable to respond to any changes in the supply rate of carbon and nutrients (Dewar, 1993; Lloyd & Farquhar, 1996), and this may prove to be more significant than is evident from the familiar mode of experimentation with plants grown in pots in fumigation chambers. Finally, at the ecosystem level any tendency of plants to grow faster or larger at elevated CO_2 will be offset to some extent by the increased demand for resources including light, nutrients and water (Zak et al., 1993; Dewar, 1993). What is required is a hierarchical modelling procedure, in which the role of the response of a biological system to CO_2 is explored at progressively greater scales (spatial extent or time period). A preliminary attempt to do this has suggested that larger scale feedbacks are much more important than small scale responses (Luan, Grace & Muetzelfeldt, 1996).

Most current experimental procedures using controlled environments do not enable a large scale approach, where for example a species of interest would form part of an established ecosystem which has endured over many species-generations. There appear to be opportunities for making such studies at naturally elevated CO_2 sites, and the following papers describe some attempts to do just this, as well as the background to some of the sites. To be sure, some new problems and questions arise when working with such natural sites; for example, fluctuations of the CO_2 concentrations (van Gardingen et al., this volume) and the extent to which the source is pure. There may be difficulty in finding plausible control areas that combine 'normal' CO_2 concentrations, that can also be regarded as similar in aspects such as vegetation and soil type. However, to be set against these drawbacks are some interesting advantages: for example the large scales of time and space, and often the presence of a distinctive isotopic signal (van Gardingen et al., 1995).

The scientific studies described in this volume (Table 1) cover three

TABLE1. Summary of research activities involving sites with naturally enhanced concentrations of atmospheric carbon dioxide described in this volume

Country	Region	Source of CO_2 emissions	Maximum CO_2 exposure[1]	Species or vegetation type	Scientific topics	Authors (*this volume*)
Italy	Tuscany (Rapolano)	Mineral spring	1000 µmol mol^{-1}	*Triticum aestivum*	Biochemistry	Badiani *et al.*
Italy	Tuscany (Bossolteo)	Mineral spring	1000 µmol mol^{-1}	*Quercus pubescens*	Carbon economy	Paoletti *et al.*
Italy	Tuscany (Bossoleto)	Mineral spring	1000 µmol mol^{-1}	*Quercus pubescens*	Litter decomposition	Ineson & Corrufo
Italy	Tuscany (Bossoleto)	Mineral spring	1000 µmol mol^{-1}	*Phragmites australis* *Quercus pubescens*	Photosynthesis Water use CO_2 environment	van Gardingen *et al.*
Italy	Campania (Solfatara)	Mineral spring	600 µmol mol^{-1}	*Agrostis canina* *Plantago major*	Growth	Fordham *et al.*
Italy	Central Italy (many sites)	Mineral springs	n/a	Grass communities	Community ecology	Selvi
Italy	Central Italy (many sites)	Mineral springs	n/a	*Scirpus lacustris*	Photosynthesis and growth	Bettarini, Miglietta & Raschi
USA	Yellowstone national park	Mineral spring	n/a	*Cyanidium caldarium*	NH_4^+ assimilation	Rigano *et al.*
USA	Utah	Burning coal seam	900 µmol mol^{-1}	Shrub community	Water use efficiency	Ehleringer, Sandquist & Philips
USA	Florida	Mineral spring	575 µmol mol^{-1}	*Boehmeria cylindrica*	Stomatal density	Woodward & Beerling
Iceland		Mineral spring	880 µmol mol^{-1}	n/a	Site description Isotopic analysis	Cook, Oechel & Sveinbjornsson
Africa	Cameroon	Degassing of lake	n/a	n/a	Site description	Mousseau, Enoch & Sabroux

[1] Maximum CO_2 levels are normally measured during the day and may exclude periods of very high concentration observed at night (see van Gardingen *et al.*, this volume).

continents (Europe, North America and Africa) with the majority conducted in Italy, in particular the site Bossoleto in central Tuscany. Most of the sites have enhanced atmospheric CO_2 concentrations resulting from outgassing at mineral springs. The range of ecosystem and vegetation types again emphasises the range of scientific activity possible using sites with naturally enhanced atmospheric CO_2 concentrations. There can be no doubt that CO_2 mineral springs will continue to be important in the scientific investigation of the effects of long-term exposure to enhanced CO_2 concentrations. The chapters in the rest of this book discuss the opportunities and potential problems using this approach, and review some of the most important work conducted at such sites to date.

REFERENCES

Bowes, G. (1993) Facing the inevitable: plants and increasing atmospheric CO_2. *Annual Review of Plant Physiology and Plant Molecular Biology*, **44**, 309-332

Dewar, R. (1993) A root-shoot partitioning model based on carbon-nitrogen-water interactions and Munch phloem flow. *Functional Ecology*, **7**, 356-368.

Emiliani, C. (1992) *Planet earth: cosmology, geology, and the evolution of life and environment*. Cambridge University Press, Cambridge.

Farquhar, G.D., von Caemmerer, S. & Berry, J.A. (1980) A biochemical model of photosynthetic CO_2 assimilation. *Planta*, **149**, 78-90.

Grace, J., Lloyd, J., McIntyre, J., Miranda, A.C., Meir, P., Miranda, H., Nobre, C., Moncrieff, J.M., Massheder, J., Malhi, Y., Wright, I.R. & Gash, J. (1995) Carbon dioxide uptake by an undisturbed tropical rain forest in South-West Amazonia 1992-1993. *Science*, **270**, 778-780.

Körner, C. (1993) CO_2 fertilization: the great uncertainty in future vegetation development. In *Vegetation Dynamics and Global Change* (eds. A.M. Solomon & H.H. Shugart), pp. 53-70. Chapman & Hall, London.

Lloyd, J., Grace, J., Miranda, A.C., Meir, P., Wong, S.C., Miranda, H.S., Wright, I., Gash, J.H.C., & McIntyre. J. (1995) A simple calibrated model of Amazon rainforest productivity based on leaf biochemical properties. *Plant, Cell & Environment*, **18**, 1129-1145.

Lloyd, J. & Farquhar, G.D. (1996) The CO_2 dependence of photosynthesis and plant growth in response to elevated atmospheric CO_2 concentration and their interrelationship with soil nutrient status. 1. General principles. *Functional Ecology* (in press).

Luan, J., Grace, J. & Muetzelfeldt, R. (1996) A hierarchical model of forest growth and succession. *Ecological Modelling* (in press).

Raven, J. (1995) The early evolution of land plants: aquatic ancestors and atmospheric interactions. Botanical Journal of Scotland, **47**, 151-175.

Van Gardingen, P.R., Grace, J., Harkness, D.D., Miglietta, F. & Raschi, A. (1995) Carbon dioxide emissions at an Italian mineral spring: measurements of average CO2 concentration and air temperature. Agricultural and Forest Meteorology, **73**, 17-27.

Wang, Y.P. & Polglase, P.J. (1995) The carbon balance in the tundra, boreal and humid tropical forests during climate change - scaling up from leaf physiology and soil carbon dynamics. *Plant, Cell and Environment*, **18**, 1226-1244.

Zalk, D.R., Pregitzer, K.S., Curtis, P.S., Teeri, J.A., Frogel, R. & Randlett, D.L. (1993) Elevated atmospheric CO$_2$ and feedback between carbon and nitrogen cycles. *Plant and Soil*, **151**, 105-117.

Migration in the ground of CO_2 and other volatile contaminants. Theory and survey

G. ETIOPE

Istituto Nazionale di Geofisica, Via di Vigna Murata 605 - 00143 Roma, Italy.

SUMMARY

Carbon dioxide is the main geosphere product affecting the biosphere. Its migration in the ground has several geological constraints whose understanding is essential for the study of its occurrence and behaviour in shallow environments, such as soil and groundwater. In industrialized, urban and rural areas high CO_2 concentrations in the ground may be associated with volatile organic compounds whose effects on the biosphere may be remarkable. In these cases any environmental monitoring should consider all the contaminants, both endogenetic and anthropogenetic, occurring in the ground. This paper is an overview of the main geological factors controlling the migration of endogenetic gas to the surface and, in particular, the CO_2 occurrence in soil and groundwater. Some guidelines are given for identifying and monitoring leakages of CO_2 and other volatile contaminants based upon soil-gas, exhalation and groundwater surveys.

INTRODUCTION

The occurrence of high concentrations of volatile compounds at the Earth's surface, viz. in soil and groundwater, is the object of a wide series of researches and practical applications within the framework of environmental studies, exploration geology and earthquake prediction. In the first case the researches are particularly focused to those gases, both endogenetic and anthropogenetic, which may have toxic relevance or excite modifications in the biosphere (e.g., CO$_2$, VOCs, CH$_4$, CFCs, SO$_2$, H$_2$S, COS, CS$_2$, HCl, HF, Rn). In the second and third case importance is given to those natural volatiles which may be used as "tracers" or "pathfinders" of subsurface energy sources (geothermal reservoirs, hydrocarbon or ore deposits), and which are sensitive to seismic stresses, respectively (e.g., He, Rn, CO$_2$, H$_2$, CH$_4$, Ar, S-compounds).

Among the subsurface gases carbon dioxide is of primary importance in many theoretical studies and bears various practical implications. The concentration of CO$_2$ in soil is an important determinant of the level of biological activity. CO$_2$ entering the groundwater is a significant component for ^{14}C dating in hydrologic studies. The rates of CO$_2$ production in soil and subsoil, and the rate at which it is transferred to the atmosphere, have an important role in global C cycling. Due to its abundance and "carrier gas" property carbon dioxide may be a "guide volatile compound" for the understanding of the mechanisms of gas migration in the ground. Finally, carbon dioxide is the object of exploitation for industrial and agricultural activity, as well as a potential hazard when associated to gas emissions linked to volcanic or geothermal activity or any geological process involving gas accumulation and sudden release to the atmosphere.

In industrialized, suburban and rural areas the study of CO$_2$ leakage in soil and subsoil is often coupled with the monitoring of VOCs (volatile organic compounds), which may seriously perturb biological activity. The sources of VOCs are many: common industrial solvents such as trichloroethylene, tetrachloroethane and benzene have been found in widespread areas. Groundwater in suburban areas has high levels of nitrates due to the use of lawn fertilizers as well as septic tank discharges. Agricultural areas have not only high levels of fertilizers but specialized synthetic organic chemicals as well. In urban and rural areas landfills and underground storage tanks holding petroleum products and synthetic organic chemicals are known sources of heavy soil and groundwater contamination. For details on VOC pollution the reader should refer to specialized literature (e.g., Mercer & Cohen, 1990; and references therein). It is worth noting that in those areas mentioned above, the study of the CO_2 effects on biological activity is not complete if it does not include surveys for locating the other contaminants affecting the biosphere and their chemico-physical interactions with CO_2.

CO_2 UNDERGROUND SOURCES

Carbon dioxide is present in low concentrations in the atmosphere (about 0.035 % v/v) but is more abundant below the ground surface due to various geochemical and biochemical processes. Owing to the easy linking between carbon and oxygen and the terrestrial abundance of these elements, carbon dioxide occurring in the shallow environments, i.e. soil and groundwater, may have various origins. Organic origin is restricted to the shallowest part of the crust where plant respiration and bacterial activity (decomposition of organic matter) occur. Carbon dioxide in the soil air is mainly the product of microbially mediated processes. Inorganic origin of CO_2 may be both shallow and deep seated. A large amount of CO_2 derives from dissolution of carbonates by meteoric water and acids produced by sulphur oxidation. CO_2 may also be produced from the oxidation of anoxic C where deep groundwaters meet shallow oxic groundwaters. Along active plate margins high temperature reactions (geothermal and magmatic processes) are a source of CO_2 at several levels of the crust. In particular, CO_2-rich waters are often produced by active metamorphism of carbonate-bearing rocks at depth. Finally, significant amounts of CO_2 at the surface may derive directly from deep seated fluid reservoirs and mantle degassing.

PROCESSES OF GAS MIGRATION

Migration in subsoil

The migration of endogenetic gases through geological formations to the surface and the interpretation of data obtained from measurements in soil air or groundwater are widely debated. The main discussion concerns the relative importance of the migration components which are: diffusion in water, diffusion in dry rocks, advection as free gas-phase (single phase flow in dry media and microbubble flow in saturated media) and advective transport by flowing groundwater (gas being in solution). Diffusion is a slow movement produced by a concentration gradient and it is described by Fick's law. If the diffusive flow is upward in the z-direction, $f_d = -D$ (dC/dz), where f_d is the diffusion flux, D is the bulk diffusion coefficient (in water or air) and C is the concentration of gas (in water or air occurring in the interstitial space of soil or rock). In general diffusion is the dominant mechanism in the intergranular channels, capillaries and small pores. In the larger pores and fractures advection is more important. For ascending gas in the z-direction, $f_a = -C\,k\,dP/\mu\,dz$, where f_a is the advective flux, the factor $(k\,dP/\mu\,dz)$ is the velocity of gas (Darcy's law, with k soil permeability and μ dynamic viscosity of interstitial gas) and C its concentration. The pressure gradient dP/dz driving the gas phase may have various origins: pressure shocks in the geological formations can be due to tectonic stresses, variation of lithostatic load, rock fracturing, localized increasing gas pressure because of overproduction of gas, charging and discharging of aquifers or crustal reservoirs. Many studies pointed out that in fractured rocks advection is dominant and that tectonic faults may be preferential channelways for degassing. Within this framework carbon dioxide plays a key role. In fact since it is one of the most abundant subsurface gases it can form "gaseous domains" (Gold & Soter, 1985), advect and carry other volatiles occurring at low amounts (e.g. noble gases, hydrogen). Several authors suggest that generally along faults, in the subsurface, a microflux of gas exists (known as "geogas"; Kristiansson & Malmqvist, 1982; Durrance & Gregory, 1990; Etiope, 1995; Etiope & Lombardi,1995); such a flux is a mixture of carrier gases (mainly CO_2) and trace gases (e.g. He and Rn). From that, it derives that high concentrations of geogas components may occur in soil and groundwater in correspondence with gas-bearing faults. As outlined by Torgersen (1990), CO_2 plays a key role in many deep seated geological processes and the quantities and fluxes of gas required to complete metamorphic and mineralogic reactions cannot be supplied by in situ sources or by diffusive transport processes. The author states that many studies provide substantial indications of the existence, persistence and the pervasive nature of enhanced advective CO_2 transport in the Earth's crust.

The CO_2 outflow at the surface may be weak, bearing concentrations in soil from the background value (i.e. biological activity) to 5-10 % v/v, or strong, with concentrations above 20-30 % or even with direct evidence of outflow (e.g. gas vents). The mean CO_2 flux at the soil-atmosphere interface is of the order of 10^{-7} - 10^{-8} m^3 m^{-2} s^{-1} (about 600-6000 t $km^{-2}y^{-1}$; Kanemasu, Powers & Sij, 1974). However, higher values can be detected over faults. The highest levels of CO_2 in soil and groundwater, often associated with gas vents, are typical of those areas characterized by hydrothermal circulation. Pressurized gas caps, linked to geothermal reservoirs or magmatic chambers (e.g. in volcanic areas), may give rise to strong surface emissions with visible effects on the surrounding environment.

Migration in soil

The movement of CO_2 and other gases in soil can be both diffusive and advective. In both cases migration from soil to the atmosphere is termed exhalation (which should not be confused with "emanation", which is the release of radon atoms from solid mineral grains to the air-filled pores). For a given gas, diffusion is controlled by the difference of the gas concentration between soil-air and atmosphere. Advective movement occurs when gas in soil has a total pressure above atmospheric. This is generally due to exogenous factors, e.g. pumping effects induced by barometric pressure changes, wind velocity or biologic overproduction of volatiles in soil. In particular geologic conditions the leakage of endogenous gas through a fault can lead to anomalous pressure and consequently advective exhalation. Another important advective process is density-driven flow. For a given gas this flow is a function of the density difference between that gas and soil-air. Density-driven gas flow is significant where the total gas density exceeds the ambient gas density by > 10% and the gas-phase permeability exceeds 10^{-11} m^2 in homogeneous media (Falta *et al.*, 1989). Even a thermal gradient between soil-air and atmosphere can lead to convective movements (namely advective movements) as is the case for vapour exhalated from the ground. It should be noted that vapour (H$_2$O), with CO$_2$, can be a "carrier gas" for volatiles occurring in trace amounts in the soil (noble gases, hydrogen, methane, VOCs, etc.) so that advective exhalation must consider these as well. For all types of migration mechanisms the gas movement patterns will be strongly influenced by soil heterogeneities which may lead to secondary phenomena (e.g. channelling, viscous fingering, dispersion).

IDENTIFYING AND STUDYING GAS LEAKAGES

Anomalous concentrations of endogenetic gases, such as CO$_2$ and H$_2$S, or anthropogenetic contaminants, such as VOCs (e.g. benzene, toluene, chloroform), in soil and groundwater may constitute a potential hazard for human health, as well as cause modification to the biological activity level. The principles of detection techniques here discussed are valid for CO$_2$ as well as for all endogenetic and anthropogenetic volatiles. They may be used to understand the contamination problems and to evaluate remediation.

Unsaturated zone

Gas anomalies occurring in the unsaturated zone can be detected and studied by monitoring wells and soil borings. In particular, gas monitoring wells are useful to detect the amount and the movement of gas, such as carbon dioxide, methane and VOCs from landfills or fuel storage tanks. However these are conventional technologies that allow monitoring in fixed points only. The soil-gas method and the exhalation measurement are quicker and less expensive techniques suitable for both reconnaissance and detailed surveys throughout the studied area. They are applicable to a wide range of volatiles under a variety of geologic and hydrologic settings. The soil-gas method is currently used as an effective tool in geology for detection of buried structures with enhanced permeability (e.g. faults;

Gregory & Durrance, 1985; Varley & Flowers, 1993; Etiope & Lombardi, 1995), ore deposits (e.g. uranium and S-minerals; Dick, 1976; Ball *et al.*, 1990), geothermal fluids (Roberts *et al.*, 1975; Vakin & Lyalin, 1990) and hydrocarbon reservoirs (Philp & Crisp, 1982; McCarthy & Reimer, 1986). The soil-gas technique has been also applied for environmental monitoring, mainly for studying pollution by VOCs in the unsaturated zone (e.g. Kerfoot, 1987). Soil-gas analysis generally involves a grab sampling which gives an instantaneous picture of the soil atmosphere at a particular subsurface location and depth, usually from 0.5 m to 1.5 m (Fig. 1).

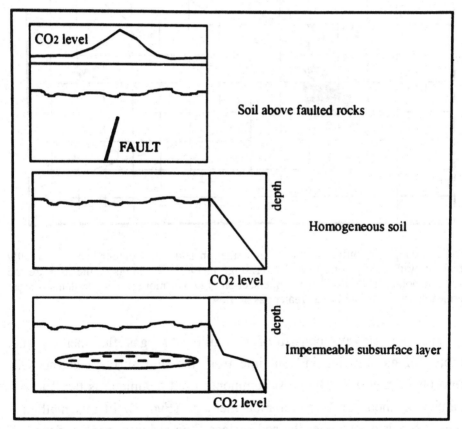

FIG. 1. CO_2 soil-gas concentrations under different soil and subsoil conditions.

The soil-gas sampling system usually consists of a portable stainless steel probe which is inserted into the ground by means of a sliding hammer.

Soil-air is extracted by a hand pump or a syringe. Analyses of gas can be performed on site, by using portable analysers (gas chromatographs, infrared detectors) or in the lab after storing gas samples in special cylinders. A complete survey can involve hundreds of sampling points and data may be computer processed and be represented as contour line charts, in which plumes or anomalous lineaments are evidenced (Fig. 2).

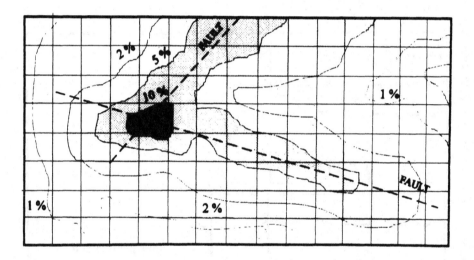

FIG. 2. Example of fault-linked CO₂ anomalies in soil-gas. Contour lines are usually obtained by the computer processed "kriging" method. The knots of the grid are the sampling points. The subsoil at the intersection of two or more fracture systems is more permeable and therefore has a greater gas outflow.

Exhalation surveys consist of measuring the gas flux density (i.e. amount of gas flowing in unit time from the unit surface) at the soil-atmosphere interface. The most commonly used technique is the "closed-chamber method" or "accumulation method" (Fig. 3). In this method a metallic box is placed over the soil surface. The efflux of gas from the soil is then proportional to the increase in gas concentration within the box over time. Gas samples are typically extracted from a sample port with a syringe for later analysis. The gas in the chamber must be well mixed, for example, by fans. Geometry and size of the chamber vary depending on the gaseous species to be measured and the purposes of the research. The flux value for a given gas is obtained by using the equation:

$$f = \frac{(C_1 - C_0) \cdot V}{\Delta t \cdot A}$$

where C_0 and C_1 are the gas concentrations at time t_0 (just after placing the chamber on the soil) and at time t_1 (usually 1-4 hours after), V is the internal volume of the box, A the area of the soil exposed to the chamber and $\Delta t = t_1 - t_0$. This formula is rigorously valid at the beginning of the accumulation, i.e. before any build-up of pressure or high concentration (inducing back-diffusion) in the chamber; errors increase as Δt increases. Theories on effectiveness and bias of this method are very well discussed by Hutchinson & Livingston (1993) and experimental tests are presented by Matthias, Blackmer & Bremner (1980).

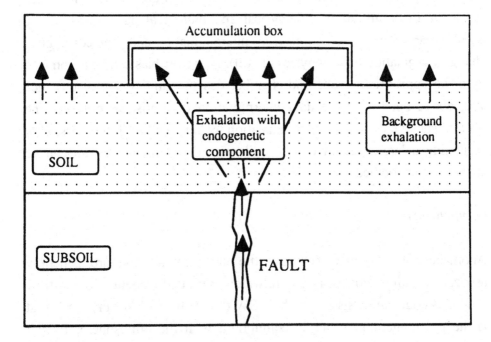

FIG. 3. Sketch of gas outflow from fractured subsoil and accumulation in the closed box.

Exhalation measurements by several methods have been performed mainly for carbon dioxide, radon and nitrogen compounds within the framework of different research fields, such as volcanic monitoring (e.g. Gurrieri & Valenza, 1988), environmental radiation protection (e.g. Wilkening, Clements & Stanley, 1972), pedologic studies (e.g. Kanemasu *et al.* 1974), and soil-air balance modelling (e.g. Kinzig & Socolow, 1994). An application to the search for gas (CO_2, Rn, He) outflow in non-volcanic areas is reported in Etiope (1995).

The comparison of soil-gas data (concentrations) with exhalation data (fluxes), referred to the same site and period, can give useful information on the behaviour of CO_2, VOCs, S-compounds and Rn in soil. Advective and diffusive movements can be identified and the potential of the gas source assessed. Moreover specific relationships with geological, atmospheric and biological factors can be studied integrating the gas surveys with geological and structural prospecting, monitoring of microclimatic and meteorological parameters, and biologic analyses. The measurements of soil-gas concentration and exhalation flux are always subject to potential sources of bias and random variability. The main bias results from perturbations of physical and biological parameters of the unsaturated zone (soil moisture, microclimatic and meteorological changes, vegetation, microphysics of soil). However bias can be minimized by using appropriate probe and chamber design, limiting soil disturbance and, above all, basing the data interpretation on statistical criteria. The main advantage of soil-gas and exhalation techniques lies in the possibility of getting a large number of data, allowing that punctual errors can be statistically smoothed.

Groundwater

Anomalies of CO_2 and VOCs in the unsaturated zone are generally associated with anomalies in groundwater. Here endogenetic CO_2 outflow produces concentrations above the ASW (Air Saturated Water) level, that is the gas content in water in equilibrium with the atmosphere. Excess CO_2, however, may be normally found in waters flowing in C-rich aquifers. Anomalies of CO_2 due to an ascending gas-phase flow can be marked by identifying in the gas mixtures high concentrations of "tracer volatiles", such as He and Rn, and by isotopic analysis of ^{13}C and ^{14}C. Anomalies of CO_2, associated with CH_4 and VOCs, often occur in correspondence with landfills, as a result of gas and leachate (a liquid resulting from mixing between rainfall and liquids occurring in the wastes) spreading downward. If the groundwater flow is high, significant concentrations of CO_2 and VOCs may occur even far from the natural or anthropogenetic sources (landfills, plants, underground storage tanks).

Groundwater may be sampled from wells and natural springs. Detailed studies on gas distribution in complex aquifers may be peformed by using multilevel sampling piezometers. In all cases collection of water samples must not be affected by atmospheric contamination. Water standing in the wells or piezometers must be removed prior to sampling. However recommendations for adequate flushing are often arbitrary and occasionally based on flawed experiments. Changes in the pressure of the sample while it is transported to the surface can cause loss of dissolved gases. These facts can introduce significant negative bias in the final data and lead to a misinterpretation of the real subsurface conditions. Such considerations limit the usefulness of several sampling devices, such as suction pumps, peristaltic pumps and air-lift pumps. Bailers and bladder pumps are considered the best sampling devices. Bailers are tubes with a check valve on the bottom; they are inexpensive and simple to operate as they can be lowered into the well on a wire or cord and yield a representative water sample. Bladder pumps are positive-displacement devices that use a pulse of gas to push the sample to the surface; the gas does not come into contact with the water sample and positive pressure is maintained at all times. The bladder pump is superior in performance to the bailer but costs more than an order of magnitude more. For water sampling at a specific depth modified medical syringes can be used (Nielsen & Yeates, 1985). Less precise, but indicative, gas data may be obtained from water supply wells with submerged pumps. In this case, however, the supply tubing should be as short and simple as possible, without intermediate containers or tanks. For shallow wells and springs hand pumps may assure a good sampling performance. In all cases water samples may be stored in glass bottles with gas-tight valves or special containers for later analysis.

Finally, extraction of dissolved gas from water can be performed from sampler headspace, after shaking. Analysis of the headspace gas can give the relative composition only, that is the percentage of the several gaseous components. To know the concentration of the several gases, that is the amount of gas dissolved in the unit volume of water, it is necessary to measure the pressure of the gas extracted within a certain volume (i.e. the number of moles); it may be carefully obtained using special extraction lines equipped with pressure gauges (e.g. Marty et al., 1991).

CONCLUDING REMARKS

The carbon dioxide outflow has several geological and physical constraints whose understanding is essential for the study of its occurrence in the biosphere. High concentrations of endogenetic CO$_2$ in soil or groundwater may be linked to several geological processes but the most important mechanism of gas emission is basically a fracture-linked advection (pressure or density driven flow), diffusion being important at low scale and in fine porous media only. Therefore the search for concealed CO$_2$ sources, or the study of CO$_2$ behaviour in soil and subsoil, cannot exclude geological and structural surveys. Exchange of endogenetic and anthropogenetic gases between subsoil, soil and atmosphere is a complex phenomenon whose measurement, by soil-gas method, exhalation method or groundwater analysis, is subject to many potential sources of bias and random variability. However, by following a carefully designed survey protocol, errors can be minimized and gas leakage can be effectively detected and gas-soil-atmosphere interactions studied over large areas and short time periods for a relatively low cost.

REFERENCES

Ball, T.K., Crow, M.J., Laffoley, N., Piper, D. & Ridgway, J. (1990) Application of soil-gas geochemistry to mineral exploration in Africa. *Journal of Geochemical Exploration*, **38**, 103-115.

Dick, W. (1976) The use of helium in mineral exploration. *Journal of Geochemical Exploration*, **5**, 3-20.

Durrance, E.M., & Gregory, R.G. (1990) Helium and radon transport mechanisms in hydrothermal circulation systems of southwest England. In *Geochemistry of gaseous elements and compounds,* pp. 337-352, Theophrastus Pub. S.A., Athens.

Etiope, G. (1995): Migrazione e comportamento del "Geogas" in bacini argillosi. *Ph.D. Thesis, Dept. Earth Sciences*, University of Rome "La Sapienza" (with extended abtract in English).

Etiope, G. & Lombardi, S. (1995) Evidence for radon transport by carrier gas through faulted clays in Italy. *Journal of Radioanalytical and Nuclear Chemistry*, **193**, 291-300.

Falta, R.W., Javandel, I., Pruess, K. & Whiterspoon, P.A. (1989) Density-driven flow of gas in the unsaturated zone due to evaporation of volatile organic chemicals. *Water Resource Research*, **25**, 2159-2169.

Gold, T. & Soter, S. (1985) Fluid ascent through the solid lithosphere and its relation to earthquakes. *Pageoph*, **122**, 492-530.

Gregory, R.G. & Durrance, E.M. (1985) Helium, carbon dioxide and oxygen soil gases: small-scale variations over fractured ground. *Journal of Geochemical Exploration*, **24**, 29-49.

Gurrieri, S. & Valenza M. (1988) Gas transport in natural porous mediums: a method for measuring CO_2 flows from the ground in volcanic and geothermal areas. *Rendiconti Società Italiana di Mineralogia e Petrografia*, **43**, 1151-1158.

Hutchinson, G.L. & Livingston, G.P. (1993) Use of chamber systems to measure trace gas fluxes. In *Agricultural ecosystem effects on trace gases and global climate change*, pp. 63-75, ASA Special Publication n° 55.

Kanemasu, E.T., Powers, W.L., Sij, J.W. (1974) Field chamber measurements of CO_2 flux from soil surface. *Soil Science*, **118**, 233-237.

Kerfoot, H.B. (1987) Soil-gas measurement for detection of groundwater contamination by volatile organic compounds. *Environmental Science Technology*, **21**, 1022-1024.

Kinzig, A.P., Socolow, R.H. (1994) Human impacts on the Nitrogen Cycle. *Physics Today*, **47**, 24-31.

Kristiansson, K. & Malmqvist, L. (1982) Evidence for nondiffusive transport of ^{222}Rn in the ground and a new physical model for the transport. *Geophysics*, **47**, 1444-1452.

Marty, B., Gunnlaugsson, E., Jambon, A., Oskarsson, N., Ozima, M., Pineau, F. &

Torssander P. (1991) Gas geochemistry of geothermal fluids, the Hengill area, southwest rift zone of Iceland. *Chemical Geology*, **91**, 207-225.

Matthias, A.D., Blackmer, A.M. & Bremner, J.M. (1980) A simple chamber technique for field measurements of emissions of nitrous oxide from soils. *Journal of Environmental Quality*, **9**, 251-256.

McCarthy, J.H.Jr., & Reimer, G.M. (1986) Advances in soil gas geochemical exploration for natural resources: some current examples and practices. *Journal of Geophysical Research*, **91**, 12327-12338.

Mercer, J.W. & Cohen, R.M. (1990) A review of immiscible fluids in the subsurface: properties, models, characterization and remediation. *Journal of Contaminant Hydrology*, **6**, 107-163.

Nielsen, D.M. & Yeates, G.L. (1985) A comparison of sampling mechanisms available for small-diameter ground water monitoring wells. *Ground Water Monitoring Review*, **5**, 83-99.

Philp, R.P. & Crisp, P.T. (1982) Surface geochemical methods used for oil and gas prospecting. A review. *Journal of Geochemical Exploration*, **17**, 1-34.

Roberts, A.A., Friedman, I. Donovan, T.J. & Denton, E.H. (1975) Helium survey, a possible technique for locating geothermal reservoirs. *Geophysical Research Letters*, **2**, 209-210.

Torgersen, T. (1990) Crustal-scale fluid transport: magnitude and mechanisms. *Eos*, **71**, 1-4-13.

Vakin, E.K. & Lyalin, G.N. (1990) Soil gas anomalies and the detection of water-conducting zones within geothermal systems. In *Geochemistry of gaseous elements and compounds*, pp. 223-242, Theophrastus Pub. S.A., Athens.

Varley, N.R. & Flowers, A.G. (1993) Radon in soil gas and its relationship with some major faults of SW England. *Environmental Geochemistry & Health*, **15**, 145-151.

Wilkening, M.H., Clements, W.E. & Stanley, D. (1972) Radon-222 flux measurements in widely separated regions. In *The natural radiation environment* (eds. J.A.S. Adams, M.W. Lowder & T.F. Gesell), pp. 717-730, USAEC, Springfield, VA.

Levels of CO$_2$ leakage in relation to geology

G. ETIOPE[1] AND S. LOMBARDI[2]

[1] *Istituto Nazionale di Geofisica, Via di Vigna Murata, 605 - 00143 Roma, Italy.*
[2] *Dipartimento di Scienze della Terra, Università "La Sapienza", P.le A. Moro, 5 - 00185 Roma, Italy.*

SUMMARY

Exhalation flux of CO$_2$ was measured by the accumulation method within a sedimentary basin in central Italy (Siena Basin). The area is characterized by underground pressurized gas linked to low enthalpy geothermal systems. The exhalation level of CO$_2$ appeared to be controlled by the fracturing (i.e. tectonization) degree of subsoil and was consistent with the distribution of high concentrations of CO$_2$, Rn and He in the soil air. The CO$_2$ flux from the whole area (about 200 km^2) would be of the order of 10 Mt y^{-1}, about one order of magnitude higher than the mean flux normally related to the soil respiration. This excess flux has clearly an endogenous component, related to underground gas domains and deep faults, acting as preferential pathways for outgassing. The results show how geologic factors influence the CO$_2$ level at the surface and how exhalation surveys, by identifying invisible surface CO$_2$ leakage, are useful for environmental characterization.

INTRODUCTION

High flows of CO$_2$ from subsurface are generally associated with volcanic activity or hydrothermal circulation. In these conditions the CO$_2$ discharge at the surface is commonly evidenced by punctual manifestations such as gas vents. In these cases it is easy to sample the gas-phase and to study its variation over time or its effects on the surrounding environment.

However, in most cases CO_2 leakage does not produce visible effects. Even in non-volcanic areas high CO_2 leakage in soil may be continuous or intermittent and generally pervasive (e.g. Hermansson et al., 1991; Etiope & Lombardi, 1995), bearing profound effects on the level of the biological activity in large areas. Identifying and studying such leakages may also be important in the search for concealed fluid reservoirs of economic importance. Such investigations can be performed by means of soil-gas (gas concentration in the soil air) and exhalation (gas flux at the soil-atmosphere interface) surveys.

The measurement of concentrations of volatiles in the unsaturated soil has been used in various research fields, basically for identifying those geological, biological or anthropogenic sources, with economic or environmental interest, affecting the composition of the soil air (underground hydrocarbon and ore deposits, geothermal reservoirs, VOC pollution, etc.; see Klusman, 1993; Etiope & Lombardi, 1995; Tedesco, 1995; and references therein for general reviews). CO_2 exhalation measurements have been performed mainly for volcanic monitoring (e.g. Gurrieri & Valenza, 1988) and pedologic studies (e.g. Kanemasu, Powers & Sij, 1974). Flux of other volatiles (such as Rn and N_2O) was measured for radiation protection zoning (e.g. Wilkening, Clements & Stanley, 1972) and soil-air balance modelling (e.g. Kinzig & Socolow, 1994). However a literature survey reveals the lack of consensus for exhalation measurements applied to environmental pollution and geological exploration.

Soil-gas measurements provide "static" information, the concentration of a gas in the soil pores, whose magnitude may be strongly affected by variation of meteorological parameters (mainly rain-induced soil moisture, atmospheric pressure, air and soil temperature, wind speed). Exhalation measurements give "dynamic" information, the flux density, and although individual data may be strongly biased, an aerial survey based on a large number of measurements could lead to a visualization of the space variation throughout vast areas of the gas leakage to the atmosphere. The coupling of exhalation and concentration data can provide useful information on CO_2 behaviour in soil and subsoil, mark pollution sources, and define relationships between underground setting and the biological activity at the surface. In particular it is possible to discriminate true signals of endogenetic gas leakage (i.e. soil-gas anomalies linked to migration from subsoil) from gas anomalies due to bias or non-geologic reasons. Moreover it is possible to have wider information on the magnitude of contamination and on the leakage behaviour of the several pollutants.

In this paper some CO_2 exhalation data, coupled with soil-gas concentrations (partially already reported in Etiope & Lombardi, 1995), are discussed to introduce the potential of this technique for geological and environmental characterization. Beyond CO_2, radon (^{222}Rn) and helium (He) were also studied. Their data are presented here as supplementary tools for better characterizing the source and mechanism of CO_2 upflow. The sampling strategy has been planned to study the relationships between exhalation level, soil-gas anomalies and underground setting of a sedimentary basin (Siena basin, central Italy).

MATERIALS AND METHODS

Soil-gas concentration at a depth of about 0.5 m has been obtained by using portable hollow steel probes, through which gas is extracted by a syringe (Fig. 1). Such a device for soil-gas sampling has already been tested in numerous researches (e.g. McCarthy & Reimer, 1986; Etiope & Lombardi, 1994; Etiope & Lombardi, 1995). Exhalation flux has been measured by the most commonly used technique, named as "closed-chamber method" or "accumulation method". This method consists of placing a metallic box with an open bottom upon soil and deducing the flux density from the build-up of gas concentration within the box. Flux is calculated by the equation:

$$f = \frac{(C_1 - C_0) \cdot V}{\Delta t \cdot A}$$

where C_0 and C_1 are the gas concentrations at time t_0 (just after placing the chamber on the soil) and at time t_1 (usually 1-4 hours after), V is the internal volume of the box, A the area of the soil exposed to the chamber

internal volume of the box, A the area of the soil exposed to the chamber and $\Delta t = t_1 - t_0$. Principles, details and limitations of this technique are reported elsewhere (e.g. Hutchinson & Livingston, 1993; Matthias, Blackmer & Bremner, 1980).

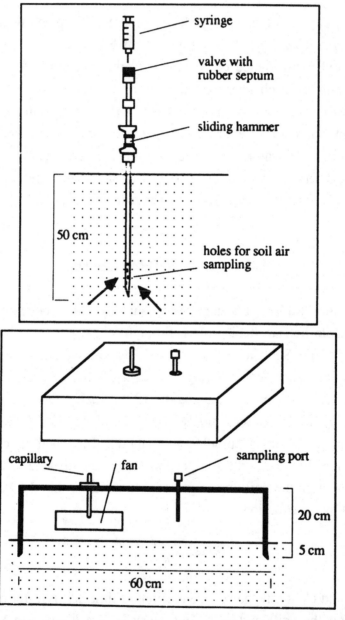

FIG. 1. Sketch of the soil-gas sampling probe (above) and the accumulation box for the measurement of gas exhalation (bottom).

The chamber used in the present work (Fig.1) has been designed in order to respect the main recommendations given by Hutchinson & Livingston (1993) to minimize bias errors linked to physical and biological disturbances (mainly temperature and pressure effects). Accordingly, the chamber consists of a square-base metallic box with side of 0.6 m and height of 0.25 m. The roof of the box is equipped with a rubber-septum sampling port and a pressure equalizing capillary. An internal fan assures mixing of air in the chamber before sampling. Then, care was taken in placing the chamber on the soil to minimize plant respiration and solar heating. Gas samples collected from the soil-gas probes and from the accumulation chamber have been analysed for CO_2, Rn and He, by gas chromatography (Sigma 3B, Perkin Elmer, Norwalk, CT, USA), by a portable scintillation counter (EDA RD 200, Eda Instruments, Toronto, Canada) and by mass spectrometry (Varian Helium Leak Detector 938-41, Varian Associates, S. Clara, CA, USA), respectively. Surveys were carried out in selected sites throughout the northern sector of the Siena Basin (central Italy), a tectonic depression filled with neogenic sands and clays. The main tectonic discontinuity and fracture zones bord the basin on the eastern side (Rapolano Fault). This area is also characterized by gas and thermal water springs linked to deep seated low enthalpy geothermal fluids with high pressure gas (CO_2) domains. A further important tectonic lineament is the "Arbia Line" which crosses transversally the northern sector of the basin and therefore cuts hundreds of metres of clay.

A soil-gas survey (249 samples) was performed with a sampling density of 1 sample per square kilometre (typical of reconnaissance surveys) throughout the northern sector of the basin. A total of 35 exhalation measurements were made in correspondence with the "Rapolano Fault", the "Arbia Line" and, for comparison, inside the basin, far from any fracture zone. Thus, three survey zones were marked (Fig. 2):

- Rapolano Fault sector (RF: 11 measurements);

- Arbia Line sector (AL: 12 measurements);

- inner basin (IB: 12 measurements)

In each measurement station, for a direct comparison between concentration and flux, two soil-gas samples were collected at t_0 and t_1 respectively and the arithmetic mean of the two concentrations obtained was considered.

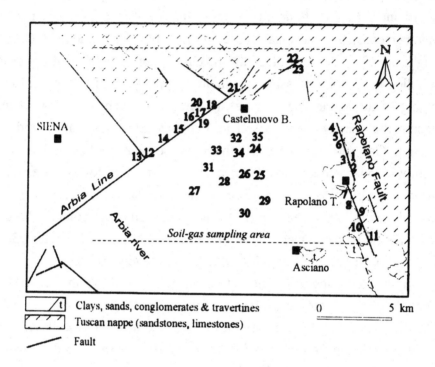

1. Bagni Freddi (a)	12. Acqua Borra (a)
2. Bagni Freddi (b)	13. Acqua Borra (b)
3. Bagni Freddi (c)	14. Monteapertaccio
4. Pod.Castiglione (a)	15. Malena-Pialli
5. Pod.Castiglione (b)	16. Pod.S.Antonio
6. Pod.Casalino	17. Pieve Pacina (a)
7. Terme S.Giovanni (a)	18. Pieve Pacina (b)
8. Terme S.Giovanni (b)	19. Pieve Pacina (c)
9. Rapolano south	20. Poggio Bonelli
10. Serre di Rapolano	21. Castelnuovo B.ga north
11. Cava Paradiso	22. Torrente Ambra (a)
	23. Torrente Ambra (b)

24. Torre a Castello
25. Monte S.Marie north
26. Monte S.Marie NW
27. Mucigliani
28. Vescona
29. La Pievina
30. Poggio S.Francesca
31. Mucigliani north
32. Castelnuovo B.ga south
33. Castelnuovo B.ga scalo
34. Castelnuovo B.ga road
35. Colonna del Grillo

FIG. 2. Sites of exhalation measurement and soil-gas survey area in the Siena Basin. Sites 1-11 are located along the eastern margin of the basin (Rapolano fault); sites 12-23 are along a fault system crossing the clayey basin (Arbia Line); sites 24-35 are located far from any fractured area.

RESULTS

Soil-gas concentration

In Table 1 some statistical data of soil-gas concentration are listed. Helium data are expressed as difference between the soil-air content and the atmospheric one (which is 5220 ppb, a value which is considered constant over time and throughout the world; Holland & Emerson, 1990).

TABLE 1. Descriptive statistical data of soil-gas concentrations.

Gas	mean	std.dev	min.val.	max.val.	samples
CO_2 (%)	1.31	3.22	0.03	46	249
^{222}Rn (Bq/l)	20	15.93	0	90.3	249
DHe (ppb)	8	163	-632	626	249
CO_2 (%) *	1.13	1.53	0.03	15.89	248

*   Data computed discarding the maximum value (46%) found close to a thermal spring

Distribution of values above the mean plus 1/2 standard deviation are highlighted on contour line maps obtained by the computer processed "kriging" interpolation method (Fig. 3a).

The main results can be summarized as follows:

- most of the higher CO_2 soil-gas concentrations lie along the "Rapolano Fault", in the eastern edge of the basin, and along the "Arbia Line", which crosses transversally the basin. In these zones CO_2 concentrations reached values up to three orders of magnitude above atmospheric.

- CO_2 and Rn anomalies showed similar distribution. Rn-CO_2 coupling evidence appears even by point-to-point analysis, as already discussed in Etiope & Lombardi (1995).

- Previous works (Etiope, 1995; Etiope & Lombardi, 1995) stressed that Rn values at some points are greatly above the maximum levels that may be supported by the very low U content of the clays (only 2-3 ppm). It has been suggested that, unless there is occurrence of localized enrichment of ^{226}Ra in soil, only an advective gas upflow from subsoil can account for the observed Rn anomalies.

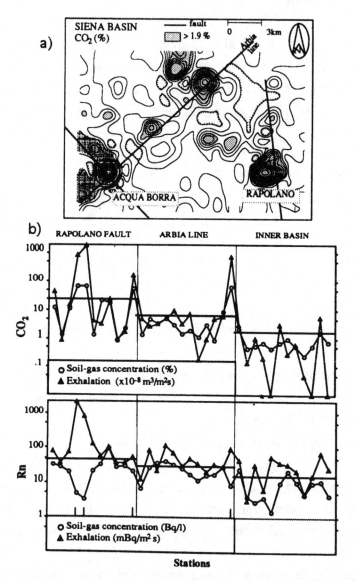

FIG. 3. a) Distribution of CO_2 in soil-gas in the northern sector of Siena Basin. b) Comparison between exhalation and concentration in soil air for CO_2 and Rn. Gas vents are marked by vertical segments. Horizontal lines show the mean exhalation level computed excluding the outflow in the gas vent areas. The magnitude of exhalation of CO_2 and Rn is congruent with the degree of the subsoil fracturing: high flux over the Rapolano Fault, medium flux over the Arbia Line and low flux far from faulted zones (inner basin).

- In faulted zones, where CO_2 is high, He concentrations reach levels greatly above the atmospheric; this only occasionally occurs in the central sector of the basin, far from faults.

Exhalation

In Table 2 statistical descriptive parameters of flux density values, obtained as total and in each station group (i.e. Rapolano Fault, Arbia Line and Inner Basin), are listed.

TABLE 2. Descriptive statistical parameters of CO_2, Rn and He flux density derived by the accumulation method. Values in brackets refer to the mean derived excluding the measurements at the gas vents.

GAS	mean	st.dev.	min.	max.	N°
Total					
CO_2 (xE-8) m^3 m^{-2} s^{-1}	105.29	348.86	0	1812	35
^{222}Rn mBq m^{-2} s^{-1}	129.34	380.68	3	2180	35
He (xE-11) m^3 m^{-2} s^{-1}	2.92	5.21	-0.55	18.41	20
Rapolano Fault					
CO_2 (xE-8) m^3 m^{-2} s^{-1}	263.1 (11.5)	568.43	0.83	1812	11
^{222}Rn mBq m^{-2} s^{-1}	326.8 (69)	654.43	36.30	2180	11
He (xE-11) m^3 m^{-2} s^{-1}	4.03	4.15	-0.23	9.81	5
Arbia Line					
CO_2 (xE-8) m^3 m^{-2} s^{-1}	64.8 (4.3)	209.51	0.17	730	12
^{222}Rn mBq m^{-2} s^{-1}	49.2 (45.2)	32.81	12.3	113	12
He (xE-11) m^3 m^{-2} s^{-1}	2.51	6.43	-0.12	18.41	8
Inner basin					
CO_2 (xE-8) m^3 m^{-2} s^{-1}	1.13	1.73	0	5.41	12
^{222}Rn mBq m^{-2} s^{-1}	28.49	20.41	3	70.8	12
He (xE-11) m^3 m^{-2} s^{-1}	2.59	4.98	-0.55	12.91	7

In all measuring stations concentration of CO_2 and Rn in the flux chamber increased over time (i.e. $C_1 > C_0$). This did not occur for He: in 6 cases out of 14 C_1 was below C_0. This fact confirms what resulted from preliminary tests, i.e. He exhalation does not always exist or is large enough to be detected by the accumulation method. Furthermore, CO_2 and Rn concentration in soil-gas proximal to the chamber was quite constant over time (Δt = 3-4 hours) and, on the contrary, He concentration strongly varied.

Four stations (n. 4, 5, 11 and 23) were next to punctual gaseous manifestations (gas vents). In these sites all three gases showed very high exhalation; at site n. 4 gas flux was so strong that the chamber became rapidly saturated and a "back-diffusion" process was evidenced by decreasing of the large He level from t_0 to t_1 (i.e. 18014 ppb to 13136 ppb). Possibly, lower back-diffusion effects occurred in the other sites next to gas vents and thus measured flux values are underestimated. In Fig. 3b exhalation is compared with soil-gas concentration. A direct is evident correlation between flux and concentration, except for Rn samples from the gas vent stations. Here Rn concentration is relatively low compared to the exhalation; this fact could be just due to the high gas flux, basically CO_2, which rapidly transport trace gases, such as Rn, and consequently does not allow any accumulation of them in the soil pores.

Relationship with tectonic setting

The soil-gas data, as a whole, highlighted the gas-bearing feature of the fault systems bounding (Rapolano fault) or crossing (Arbia Line) the basin. However the soil-gas anomalies along the Arbia Line do not occur homogeneously throughout the fault line: anomalous points often alternate with the lack of any gas signal. Furthermore, high soil-gas concentrations, although occasionally, were detected within the central sector of the basin, far from any faulted zones. The occasional lack of leakage signals along the Arbia Line may suggest that here faults are not totally permeable to gases throughout their extension. However it is possible that some data are biased (due to variations of meteorological parameters and micro-physics of soil and to prospecting errors), i.e. leak signals are masked, as well as "false signals", produced in the unfractured zones. This phenomenon is quite common in soil-gas surveys and represents the main limitation of this technique.

The exhalation flux gives further cognitive elements on the occurrence of gas leakage from subsoil. The diagrams of Fig. 3b show that for both CO_2 and Rn the distinction of the three survey areas matches the exhalation level. Measuring stations localized within the basin, far from fault zones, display the lowest level of exhalation. In the groups of fault zones exhalation is on average higher, being the highest in the Rapolano Fault zone. In correspondence with the Rapolano Fault the clay thickness is largely reduced or absent and, however, all rocks (travertines, limestones) are heavily tectonized. The enhanced permeability condition is widely testified by spreading thermal springs (CO_2 rich). This feature is reflected in the exhalation survey (even far from punctual fluid manifestations) that evidenced a pervasive distribution of the gas leakage (as already suggested by the soil-gas survey). On the contrary, inside the basin flux values are systematically low. They reflect the normal production and exhalation of CO_2, linked to surface biologic activity, and of Rn, linked to decay of ^{226}Ra naturally occurring in soil and outcropping rocks. The exhalation level at the Arbia Line is intermediate between the Rapolano Fault one and the inner basin one. This fact seems to be justified by the geologic conditions of this sector, which from the permeability viewpoint are just intermediate between the margin and the centre of the basin: that is, the Arbia Line zone is characterized by a thick clay cover (about 400-600 m) which is crossed and consequently fractured by a tectonic discontinuity.

CONCLUSIONS

The exhalation level of CO_2 was related to the Rn one and to the fracturing degree of subsoil linked to tectonics; it may be divided into three categories:

- high exhalation level, at the Rapolano fault;
- medium exhalation level, at the Arbia Line;
- low exhalation level (background), inside the basin.

The CO_2 flux from the whole area (about 200 km^2) would be of the order of 10 Mt y^{-1}, about one order of magnitude higher than the mean flux normally occurring in the ground (about 3000-6000 t km^{-2} y^{-1}; Kanemasu *et al.,* 1974). This excess flux has clearly an endogenous component, related to underground gas domains and deep faults, acting as preferential pathways for outgassing. The occurrence of such a leakage has also been highlighted by the soil-gas survey.

A direct proportionality exists between exhalation and concentration of soil gas, except for Rn in the gas vent areas, where "flushing" phenomena (due to ascending CO_2) may prevent trace gas accumulation in soil pores. Helium exhalation is detectable only if high gas flux exists, as in correspondence with gas vents. Possibly, the accumulation method is not sensitive enough for lighter gases, such as helium, whose movement from soil to air is strongly affected by local disequilibrium in the soil-atmosphere exchanges and by the variation of micro-physics parameters in soil. Taking into account all possible imprecisions and bias, exhalation and soil-gas data of CO_2 and Rn are consistent with the geologic setting of the studied area. Both surveys pointed out the occurrence of gas leakage connected to faults, both in the margin and in the middle of the Siena Basin, regardless of the clay thickness.

Data discussed in this work, although limited, show the role of faulted rocks on the CO_2 level at the surface and are an example of how exhalation surveys, by identifying over large areas invisible surface CO_2 leakage, are useful for environmental characterization. Further volatile compounds affecting the biosphere could be studied in the same way. Then, a wider data-base, reflecting various geologic and morphologic conditions, may allow a more complete knowledge of CO_2 leakage and of its possible implications in the surrounding environment.

ACKNOWLEDGEMENTS

Thanks are due to Mr. E. Grillanda, Mrs. A. Baccani and F. Zampetti for technical support during laboratory and field work. Part of this research was supported by EC funds (contract N° FI2W-CT91-0064).

REFERENCES

Etiope, G. (1995) Migrazione e comportamento del "Geogas" in bacini argillosi. *Ph.D. Thesis, Dept. Earth Sciences*, University of Roma "La Sapienza" (with extended abstract in English).

Etiope, G. & Lombardi, S. (1994) ^{222}Rn soil-gas in sedimentary basins in central Italy: its implications in the radiation protection zoning. *Radiation Protection Dosimetry*, **56**, 231-233.

Etiope, G. & Lombardi, S. (1995) Evidence for radon transport by carrier gas through faulted clays in Italy. *Journal of Radioanalytical and Nuclear Chemistry*, **193**, 291-300.

Gurrieri, S. & Valenza M. (1988) Gas transport in natural porous mediums: a method for measuring CO_2 flows from the ground in volcanic and geothermal areas. *Rendiconti Società Italiana di Mineralogia e Petrografia*, **43**, 1151-1158.

Hermansson, P.H., Akerblom G., Chyssler, J. & Linden, A. (1991) Geogas - a carrier or a tracer? In *Statens Karnbransle Namnd (SKN) Report 51*, pp.66. National Board for spent nuclear fuel, Stockholm.

Holland, P.W. & Emerson, D.E. (1990) The global helium-4 content of near-surface atmospheric air. In *Geochemistry of gaseous elements and compounds*, pp. 97-109. Theophr. Pub. S.A., Athens.

Hutchinson, G.L. & Livingston, G.P. (1993) Use of chamber systems to measure trace gas fluxes. In *Agricultural ecosystem effects on trace gases and global climate change*, pp. 63-75. ASA Special Publication n° 55.

Kanemasu, E.T., Powers, W.L. & Sij, J.W. (1974) Field chamber measurements of CO_2 flux from soil surface. *Soil Science*, **118**, 233-237.

Kinzig, A.P., Socolow, R.H. (1994) Human impacts on the Nitrogen Cycle. *Physics Today*, **47**, 24-31.

Klusman, R.W. (1993) *Soil gas and related methods for natural resource exploration.* J.Wiley & Sons, Chichester.

Matthias, A.D., Blackmer, A.M. & Bremner, J.M. (1980) A simple chamber technique for field measurements of emissions of nitrous oxide from soils. *Journal of Environmental Quality*, **9**, 251-256.

McCarthy, J.H.Jr. & Reimer, G.M. (1986) Advances in soil gas geochemical exploration for natural resources: some current examples and practices. *Journal of Geophysical Research*, **91**, 12327-12338.

Tedesco, S.A. (1995) *Surface geochemistry in petroleum exploration.* Chapman & Hall, New York.

Wilkening, M.H., Clements, W.E. & Stanley, D. (1972) Radon-222 flux measurements in widely separated regions. In *The natural radiation environment II*, (eds. J.A.S. Adams, M.W. Lowder & T.F. Gesell), pp. 717-730. USAEC, Springfield, VA

CO_2 emission in volcanic areas: case histories and hazards

M. MARTINI

Department of Earth Sciences, Università di Firenze, Italy.

SUMMARY

Carbon dioxide is a significant component of volcanic fumaroles, and its concentration can vary within large limits in different environments. A high level of hazard can be associated to natural gas emissions, and a concentration of about 30% CO_2 in the atmosphere is considered as a life threatening threshold. The oldest historical record of fatalities produced by natural degassing dates back to the Pompeii eruption of Vesuvius (AD 79); more recently, 142 people were killed in 1979 on the Dieng plateau (Indonesia), and gas emissions from volcanic lakes of Cameroon, mainly consisting of CO_2, caused 37 victims at Monoun in 1984 and over 1700 fatalities at Nyos in 1986. Significant outputs of carbonic gases can be expected at any active or dormant volcanic system, so that potential hazards from toxic gases are associated to any area of recent or present volcanism. With reference to Italy, two sites appear to deserve our attention when considering hazards connected with gaseous emissions: Phlegrean Fields, NW of Naples, and the island of Vulcano, where civil settlements are located well inside the potentially hazardous areas.

INTRODUCTION

Because of the low solubility of carbon in silicic melts, carbon oxides are preferentially released from magma bodies as the confining pressures decrease: accordingly, any volcanic activity produced by the ascent of magmas from depth is characterized by significant outputs of carbon dioxide, whose stability is much higher with respect to carbon monoxide in all natural conditions.

CO$_2$ outputs are reported for different active volcanoes; values as high as 11,500 tons/day for Mt. St. Helens (Harris *et al.*, 1981), 184,000 tons/day for Nyiragongo (Le Guern, 1982) and 120,000 tons/day for Etna (Allard *et al.*, 1991) have been estimated.

CO$_2$ concentrations in volcanic fumaroles vary within wide limits (Table 1) and no strict dependence upon tectonic settings, types of magmas or outlet temperatures can be derived; the contribution of organic material and carbonate rocks appears on the contrary to play a major role in CO$_2$ discharges (Irving & Barnes, 1980).

TABLE 1. CO$_2$ concentrations in volcanic fumaroles.

Volcano	Temp. °C	CO$_2$ % vol.	Ref.
Klyuchevskoy (Kamchatka)	1070	0.05	(1)
Ebeko (Kurile)	235	0.92	(2)
Tokachi (Japan)	339	1.59	(3)
Nevado del Ruiz (Colombia)	85	1.75	
Kilauea (Hawaii)	900	2.33	(4)
Vulcano (Italy)	570	10.60	
Teide (Canary)	89	18.64	
Phlegrean Fields (Italy)	158	21.12	
Stromboli (Italy)	410	29.68	

(1) Taran *et al.*, 1991
(2) Menyailov, Nikitina & Shapar, 1985
(3) Hirabayashi, Yoshida & Ossaka, 1990
(4) Greenland, 1984

Whatever the origin, however, CO$_2$ emissions represent a significant natural hazard, and fatalities because of excessive concentrations of this gas phase in the environment are commonly reported.

Even if CO$_2$ is not the most toxic among natural gases (Table 2), its actual concentrations can attain sufficient values to produce the death of humans and/or animals.

The values listed in Table 3, obtained by dividing the concentrations in fumarolic gases by the life threatening levels, seem to indicate that hazardous concentrations occur rarely for CO_2, and more frequently for other species like HCl, SO_2, H_2S, and HF.

TABLE 2. Life threatening levels of natural gases (Baxter & Kapila, 1989).

CO_2	30 % volume
HCl	0.1
SO_2	0.04
H_2S	0.1
CO	0.4
HF	0.015

Chlorine, sulfur, and fluorine compounds, however, are significantly soluble in aqueous solutions and are normally trapped in water vapour condensing at the fumarole outlet; on the contrary, because of its density, CO_2 can accumulate in morphologically depressed areas, attaining concentrations well above the mentioned life threatening level.

TABLE 3. Life threatening ratios for different fumarolic gases.

Volcano	CO_2	HCl	SO_2	H_2S	CO	HF
Klyuchevskoy	0.002	15	2.5	0.50	0.38	19
Ebeko	0.03	0.5	2.5	1.1	-	0.06
Tokachi	0.05	1.5	50	6.4	-	3.0
Nevado del Ruiz	0.06	1.2	41	0.17	0.001	1.2
Kilauea	0.08	2.1	280	3.9	1.9	17
Vulcano	0.33	2.0	11	2.0	0.02	2.0
Teide	0.63	-	-	0.13	0.001	-
Phlegrean Fields	0.70	0.09	-	1.5	0.001	0.01
Stromboli	0.98	2.8	38	0.10	0.001	2.1

CASE HISTORIES

Vesuvius (Italy)

The eruption of Vesuvius in AD 79 is the first ever described, and the famous letters of Pliny the Younger to the historian Tacitus give a sufficient idea of what happened on August 24 and 25 (Bullard, 1976).

Besides the different types of volcanic products which allowed a detailed reconstruction of the eruptive event, however, our consideration is mainly devoted to the postures of the gypsum casts which the excavators obtained where hollow spaces appeared to indicate the burial sites of people escaping from Pompeii (Fig. 1). Since the fall of fine pumice, which is spread all around, cannot account for the deaths of these people, suffocation because of gaseous emissions can rather be taken into account.

Even minor concentrations of HCl, SO_2, H_2S, and HF, provoke great discomfort and irritation, while CO_2 produces gradual and peaceful loss of consciousness; the quiet posture observed for the casts in Fig. 2 fully agrees with this last possibility, so that a CO_2 degassing appears as the most probable killing agent.

Dieng (Indonesia)

On February 20th, 1979 a small phreatic eruption occurred at night on the Dieng plateau (Java), producing voluminous steam emission (Le Guern, Tazieff & Faivre-Pierret, 1982).

Because of the seismic shocks which preceded and accompanied the eruptive event, 142 people living at a short distance from the volcanic centre escaped from their houses through a path entering a low-lying area; the following morning all of them were found dead because of the excessive CO_2 concentration accumulated in that site during the minor volcanic event.

Fig. 1 and 2. Gypsum casts at Pompeii, representing victims of the eruption of Vesuvius in AD 79.

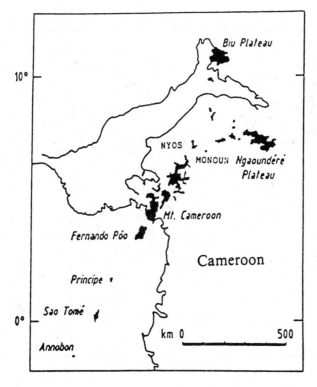

FIG. 3. Location of Lakes Monoun and Nyos (modified from Barberi *et al.*, 1989).

Monoun (Cameroon)

A burst of gases from Lake Monoun (Fig. 3) occurred on August 15th, 1984 and a whitish cloud spread around the shores, invading also the nearby road (Sigurdsson *et al.*, 1987); 37 people who stepped into the area during the following four hours were found dead. Subsequent investigation led to the conclusion that a landslide from the lake's eastern rim had caused upwelling of deep water saturated with respect to CO_2, which was suddenly released to the atmosphere.

Nyos (Cameroon)

A significant emission of gases occurred on August 21th, 1986 from the waters of Lake Nyos, located in the same volcanic area as Lake Monoun (Fig. 3).

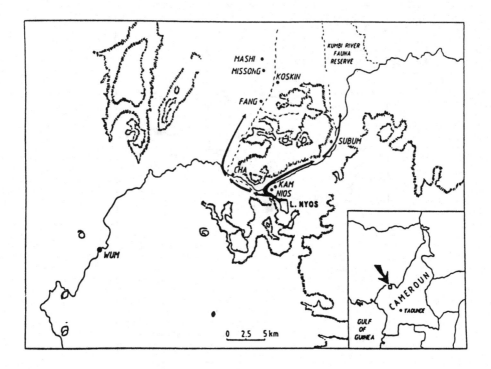

FIG. 4. The main route of the gas cloud emitted from Lake Nyos (modified from Barberi et al., 1989).

Because of the relative elevation of the lake basin with respect to the surrounding area, the gas cloud travelled downslope affecting several human settlements (Fig. 4) and producing a toll of over 1700 fatalities. Even if discordant opinions have been expressed about the trigger of the phenomenon, CO$_2$ was again indicated as the main component of the gaseous phase (Sigvaldason, 1989).

PRESENT HAZARD IN ITALY

Emission of natural gases, among which CO$_2$ normally displays the most significant concentrations, can occur in areas of recent volcanism or of tectonic activity; accordingly, some degree of hazard can be associated to active faults or to quaternary volcanic systems.

Lakes Bolsena, Vico, Bracciano, Albano, Nemi, and Averno, hosted in calderas or craters of recent eruptive activities in central Italy, and around which diffuse degassing from the ground is normally occurring, all can represent possible sites for significant CO_2 outputs. A comparison with the situation of the lakes in Cameroon is not easy; a certain degree of instability, however, is reported in the historical record for Lake Albano (Martini *et al.*, 1994).

The probability of natural gas emission is much higher in areas of active volcanism. Significant CO_2 outputs from the ground also pertain to Stromboli and Etna, but the slope acclivities of their craters as well as sufficiently strong winds prevent the accumulation of hazardous concentrations. A higher degree of concern, on the contrary, can be associated to Phlegrean Fields and the island of Vulcano, which deserves further attention.

Phlegrean Fields

The volcanic area of Phlegrean Fields, where the last eruptive event occurred in 1538, has been characterized since the Roman times by vertical movements termed as "bradyseismic", which produced uplifts and sinking phases with a relative extent of at least 10 m.

An elevation of about 0.7 m occurred in 1970-1972 and then, after a quiet stage of ten years, a new upheaval phase started in 1982 giving rise to a further elevation of 1.5 m to 1984. After this, a moderate sinking phase is still in progress.

Fumarolic gases are at present emitted inside and around the Solfatara of Pozzuoli; a mean CO_2 concentration of about 20% is observed (Martini *et al.*, 1991) and a total output of 350 tons/day has been estimated (Carapezza *et al.*, 1984).

The flat area of Solfatara is normally swept by gentle breezes, and no problem related to CO_2 accumulation has been reported; the coincidence of unfavourable natural conditions, however, can produce at any moment the build-up of risky gaseous concentrations.

Vulcano

Subsequent to the last eruption in 1888-1890, fumarolic activity of varying extent characterizes the crater La Fossa (Fig. 5). Systematic geochemical surveillance started in 1977 has shown more than negligible changes in chemical composition and total output of gaseous species. An estimation of 180 tons/day has been provided for the crater and 30 tons/day for the surrounding area (Baubron, Allard & Toutain, 1990).

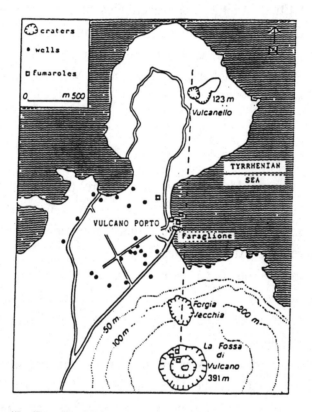

FIG. 5. Crater "La Fossa" at Vulcano.

Because of the increased contribution of heat and water to the volcanic system, an inflation in the crater area producing the opening of new fractures is in progress since 1987 (Martini, 1993).

Even if the probability of a true volcanic event is not very high, the still active deformation process can trigger a collapse in the weakest sector of the crater, facing the village of Vulcano Porto. Decompression of the deep system can follow with stronger gas emission; if no favourable weather conditions dissipated the gas cloud, the village could be invaded by noxious gases with severe consequences. If we take into account that during summer holidays the island hosts about 10,000 people mainly living in Vulcano Porto, the consequent hazard appears to exceed any prudent threshold.

REFERENCES

Allard P., Carbonnelle, J., Dajlevic, D., Le Bronesc, J., Morel, P., Robe, M.C., Maurenas, J.M., Faivre-Pierret, R., Martin, D., Sabroux, J.C. & Zettwoog, P. (1991) Eruptive and diffuse emissions of CO_2 from Mount Etna. *Nature*, 351, 37-40.

Barberi, F., Chelini, W., Marinelli, G. & Martini, M. (1989) The gas cloud of Lake Nyos (Cameroon, 1986): results of the Italian technical mission. *J. Volcanol. Geotherm. Res.*, 39, 125-134.

Baubron, J.C., Allard, P. & Toutain, J.P. (1990) Diffuse volcanic emissions of carbon dioxide from Vulcano island, Italy. *Nature*, 344, 51-53.

Baxter, P.J. & Kapila, M. (1989) Acute health impact of the gas release at Lake Nyos, Cameroon, 1986. *J. Volcanol. Geotherm. Res.*, 39, 266-275.

Bullard, F.M. (1976) Volcanoes of the Earth (Italian translation). Newton Compton, Roma, 636 pp.

Carapezza, M., Gurrieri, S., Nuccio, M.P. & Valenza, M. (1984) CO_2 and H_2S concentrations in the atmosphere at the Solfatara of Pozzuoli. *Bull. Volcanol.*, 47, 287-293.

Greenland, L.P. (1984) Gas composition of the January 1983 eruption of Kilauea volcano. *Geoch. Cosm. Acta*, 48, 193-195.

Harris, D.M., Sato, M., Casadevall, T.J., Rose, W.I. & Bornhorst, T.J. (1981) Emission rates of CO_2 from plume measurements. *USGS Prof. Paper*, 1250, 201-207.

Hirabayashi, J., Yoshida, M. & Ossaka, J. (1990) Chemistry of volcanic gases from the 62-1 crater of Mt. Tokachi, Hokkaido. Japan, *Bull. Volcanol. Soc. Japan*, 35, 205-215.

Irwin, W.P. & Barnes, I. (1980) Tectonic relations of carbon dioxide discharges and earthquakes. *J. Geophys. Res.*, 85, 3115-3121.

Le Guern, F. (1982) Les débits de CO_2 et de SO_2 volcaniques dans l'atmosphère. *Bull. Volcanol.*, 45, 197-202.

Le Guern, F., Tazieff, H. & Faivre-Pierret, R. (1982) An example of health hazard: people killed by gas during a phreatic eruption. Dieng plateau (Java, Indonesia), February 20th 1979. *Bull. Volcanol.*, **45**, 152-156.

Martini, M., Giannini, L., Buccianti, A., Prati, F., Cellini, Legittimo, P., Iozzelli, P. & Capaccioni, B. (1991) 1980-1990: ten years of geochemical investigation at Phlegrean Fields. *J. Volcanol. Geotherm. Res.*, 48, 161-171.

Martini, M. (1993) Water and fire: Vulcano island from 1977 to 1991. *Geochem. J.*, **27**, 301-307.

Martini, M., Giannini, L., Prati, F., Tassi, F., Capaccioni, B. & Iozzelli, P. (1994) Chemical characters of crater lakes in the Azores and Italy: the anomaly of Lake Albano. *Geochem. J.*, **28**, 173-184.

Menyailov, I.A., Nikitina, L.P. & Shapar, V.N. (1985) Results of geochemical monitoring of the activity of Ebeko volcano (Kurile islands) used for eruption prediction. *Geodynamics*, **3**, 259-274.

Sigurdsson, H., Devine, J.D., Tchoua, F.M., Presser, T.S., Pringle, M.K.W. & Evans, W.C. (1987) Origin of the lethal gas burst from Lake Monoun, Cameroon. *J. Volcanol. Geotherm. Res.*, **31**, 1-16.

Sigvaldason, G. (1989) International Conference on Lake Nyos Disaster, Yaoundé, Cameroon 16-20 March 1987: Conclusion and Recommendations. In *The Lake Nyos Event and Natural CO$_2$ Degassing. I* (eds. F. Le Guern & G. Sigvaldason). *J. Volcanol. Geotherm. Res.*, **39**, 97-107.

Taran, Yu. A., Rozhkov, A.M., Serafimova, E.K. & Esikov, A.D. (1991) Chemical and isotopic composition of magmatic gases from the 1988 eruption of Klyuchevskoy volcano, Kamchatka. *J. Volcanol. Geotherm. Res.*, **46**, 255-263.

Controlled degassing of lakes with high CO_2 content in Cameroon: an opportunity for ecosystem CO_2-enrichment experiments

M. MOUSSEAU[1], Z.H. ENOCH[2] AND J.C. SABROUX[3]

[1] *Laboratoire d'Ecologie Végétale, URA 1492, Université Paris-sud, 91405 Orsay Cedex, France.*
[2] *Agricultural Research Organisation, The Volcani Center, Institute of Soils and Water, Bet Dagan, Israel.*
[3] *Institut de Protection et de Sûreté Nucléaire/DPEI/SERAC, Centre d'Etudes de Saclay, 91191 Gif sur Yvette Cedex, France. Also : International Union of Technical Associations and Organizations (UATI), 1 rue Miollis, 75732 Paris Cedex 15, France.*

SUMMARY

In 1984 and 1986, massive eruptions of carbon dioxide from two lakes in Cameroon killed at least 1800 Cameroonian villagers. Countless head of cattle were also asphyxiated. Since then, measurements have shown that the amount of CO_2 still dissolved in these lakes is very high (20,000 tons and 500,000 tons at Monoun and Nyos, respectively). The danger of a future gas burst (the so-called "limnic eruption") could be eradicated by drawing off dissolved CO_2. A gas-lift experiment at Lake Monoun in April 1992 allowed CO_2 to be released at a flow-rate of 15 to 150 $l.s^{-1}$ (STP), depending on the diameter of the pipe used. The large high grade CO_2 resource from these two west African lakes provides an exceptional opportunity to conduct large scale and long-term experiments on the effect of increased atmospheric CO_2 concentrations on biotic systems in the tropical region. A description of the planned experiments is presented.

INTRODUCTION

In several volcanic regions of the world, large amounts of gas containing mainly CO_2 are sporadically released. In some instances, the gas released may asphyxiate humans. In the last decade, this phenomenon has been reported in Indonesia (Le Guern, Tazieff & Faivre-Pierret, 1982), and later in Cameroon at lake Monoun in 1984 (Sigurdsson et al., 1987) and at Lake Nyos in 1986. These gas eruptions were all of nearly pure carbon dioxide of deep (mainly magmatic) origin, expanding from a near-surface reservoir.

In the case of the Monoun and Nyos eruptions, the reservoir feeding the gas burst is most likely the lake water itself, in which huge amounts of carbon dioxide are still dissolved. Degassing the two lakes has been widely and repeatedly proposed to alleviate the risk of future gas bursts from Monoun and Nyos. Controlled expansion of the gas through vertical pipes will drive gas-rich bottom waters up to the surface in a self powered gas-lift. The safe disposal of carbon dioxide (and degassed water) is still being debated. Besides mitigating an appalling natural hazard, might this disposal also be useful for biological research?

Net primary production of natural vegetation is known to be enhanced by increased atmospheric CO_2 in the short term (Strain & Cure, 1985). Feedback limitations, however, may negate this fertilizing effect in the longer term (Bazzaz, 1990).

Very few experiments of CO_2 enrichment have been performed for a long term on natural ecosystems (Strain & Thomas, 1992): tundras (Oechel & Billings, 1992), salt marshes (Drake, 1992) or mediterranean macchia (Scarracia Mugnozza et al., 1993). To our knowledge, no CO_2 experiments have been performed on natural tropical ecosystems, except that made by Körner & Arnone (1992) with an artificially recomposed tropical ecosystem in a growth chamber.

The controlled degassing of the Cameroon lakes offers an unique opportunity for a long term CO_2 enrichment experiment on tropical plants and small plots representing the ecosystem. The potential use of the CO_2, to be released into the lake's surroundings, is presented after a brief description of the geographical, ecological and limnological settings of the lakes.

DESCRIPTION OF THE LAKES

The geographical setting

The lakes are situated in the Northwest province of Cameroon, on the "Volcanic basaltic line" that intersects a pre-Tertiary granitic plateau (see map, Fig. 1). Lake Monoun is located at an altitude of 1080 m (5°34'N, 10°35'E) and lake Nyos is 130 km to the NW, in an ancient volcanic crater at an altitude of 1091m (6°26'N, 10°18'E). Both lakes are volcanic in origin, although the shape of lake Monoun is far from the circular pattern of a typical crater lake: in this latter case, the crater rim is concealed under 10 m of water accumulated in a river bed. Lakes Monoun and Nyos are 96 m and 208 m deep, respectively.

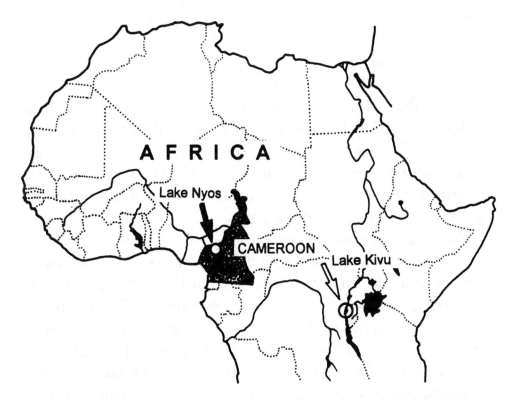

FIG. 1. Geographical setting of known gas-bearing lakes in Africa (after Stager, 1987).

The vegetation

The vegetation around lake Nyos is a bushy savanna, characteristic of medium altitude, often called periforest savanna of the guinean-soudanean zone (Le Touzey, 1985). It is primarily an herbaceous ecosystem, heavily grazed and frequently burnt, dominated by graminaceous species (Andropogonae). The species *Hyparrhenia diplandra* often represents 50% of the herbaceous flora (Le Touzey, 1985). Some small trees grow on the banks of the lake together with some woody bushes.

The lake Monoun region has the same phytosociological origin. However, the original vegetation has been heavily changed by humans and grazed by cattle. It is located in a highly cultivated area: maize, bananas and arachids are the major crops, but the village people also grow and eat a large number of tubers (such as cassava, yam, macabo or *Caulocasia* tubers).

The gas content of the lakes

The true nature of the gas disaster that struck northwest Cameroon in 1984 and 1986 is still a matter of controversy. Experts agree on the composition and origin (almost pure magmatic CO_2) of the dense gas clouds that asphyxiated all human beings and animals in their path. On the other hand, it is still unknown whether the gas was stored **under** or **within** the lakes before the catastrophic bursts.

Since the Nyos disaster of August 1986, however, considerable evidence has emerged to strengthen the second hypothesis. According to this, the deadly carbon dioxide (may be up to 1 km₃ STP in the case of Lake Nyos) was dissolved in the deep water layers of the lake prior to eruption, and was subsequently released by a lake overturn. This "limnic eruption" (Sabroux, Dubois & Doyotte, 1987) was first described by Sigurdsson *et al.* (1987) after the August 1984 eruption at lake Monoun.

Among evidences, the current gas content of the lakes ranks first. Lake Monoun and lake Nyos still contain 20,000 tons and 500,000 tons of dissolved CO_2, respectively (Tietze, 1992; Halbwachs *et al.*, in prep.). Gas is being added continuously to the lakes from depth. The maximum concentration of CO_2 is *ca* 3.5 litres (STP) and *ca.* 7.5 litres (STP) per litre of water in lake Monoun and lake Nyos respectively. As solutions of carbon dioxide in water are denser than pure water, the lakes are stratified, but this chemical stratification is offset by the thermal stratification. The delicate balance between both stratifications determines the stability of the lake. The overall stratification pattern is unpredictable, since it is caused by a double antagonistic transport process – diffusion of dissolved gas and conduction of heat – within a gravitational field (dissipative structure). Although CO_2 and heat fluxes may be assumed constant, gas bursts are by no means periodical.

When a water layer reaches CO_2 saturation, any disturbance may cause it to erupt, releasing its formerly dissolved carbon dioxide into the atmosphere. Therefore, there is no assurance that the lakes Nyos and Monoun will remain stable for ever.

The first gas lift experiment at lake Monoun

The simplest method to alleviate the risk of future gas bursts from lakes Monoun and Nyos is to pump the gas off the lake waters by means of a self-powered gas-lift. Such a gas-lift has been operating for several years at Gisenyi, Rwanda, to extract methane (and carbon dioxide) from lake Kivu.

An experiment was done at lake Monoun in 1992 (Halbwachs *et al.*, 1993). Two vertical pipes of different diameter were set up vertically between the lake bed and the surface. Once the gas-lift is primed in a pipe by using a mechanical pump to raise the water in the pipe, to the level where bubbles nucleate, it is self-sustaining by the expansion of the gas bubbles above the saturation level. The result is a steady jet of a gas-water mixture at the pipe head. The diphasic fluid thus discharged is replaced by gas-rich water at the pipe bottom, as long as gas is available for expansion.

With a 53 mm inner-diameter pipe, the flow rate was $4.6 \, l \, s^{-1}$ of water and $15 \, l \, s^{-1}$ of carbon dioxide. With one 141 mm inner-diameter pipe, the figures rose to $44 \, l.s^{-1}$ and $151 \, l \, s^{-1}$, respectively, with the mechanical power of the jet as high as 3.4 kW. Two to three larger diameter pipes could easily remove the carbon dioxide currently dissolved in lake Monoun in one year.

WHAT CAN BE DONE WITH THE CO_2 RELEASED?

During a workshop held in Paris in 1992[1], degassing of the lakes was highly recommended and the idea arose that the released CO_2 could be used to conduct large scale experiments on the response of biotic systems to elevated atmospheric CO_2. We propose to investigate the response of a tropical ecosystem to global elevated CO_2 by linking the degassing of the lakes to an experimental design using the water and the extracted CO_2. We emphasize here that the carbon dioxide from lakes Monoun and Nyos is "high grade" for this purpose: it is virtually free of ethylene (C_2H_4) and hydrogen-sulphide, which could hinder plant metabolism (Table 1). The absence of C_2 to C_4 hydrocarbons suggests that CH_4 inputs in these lakes are of biogenic origin (Kling, Evans & Tuttle, 1991).

TABLE 1. Gas composition in ppmv. Data from Kling *et al.* (1991), Giggenbach (1990) and Sigurdsson *et al.* (1987).

	CO_2	CH_4	N_2	O_2	He	H_2S	C_2H_4
MONOUN	986,400	12,620	995	<20	5	<200	<0.1
NYOS	998,100	1380	517	20	3	<10	<0.002

We suggest 3 types of investigations:

1) *Survey of the present CO_2 atmospheric concentrations and environmental parameters*

Wind speed, temperature, rainfall and atmospheric humidity should be monitored in the neighbourhood of the lakes. A meteorological station and CO_2 analysers should be set up at the shore of the lakes to obtain preliminary data on environmental parameters and their temporal and spatial variation. It is likely that the CO_2 concentration at some locations will be above the global average. The morphology and physiology of plants growing at those locations should be studied. The meteorological

[1]*Dégazage controlé des lacs Monoun et Nyos*. Paris, 9-11 Septembre 1992, UATI-FMOI and Fondation Gaz de France.

and CO_2 stations should be kept operating throughout the whole degassing process.

2) Survey of historical CO_2 concentrations

CO_2 released from the lakes is of magmatic origin. It thus contains virtually no ^{14}C and has a $\delta^{13}C$ different from the $-8^0/_{00}$ PDB typical of atmospheric CO_2 (Table 2).

Carbon isotope ratios in live and dead plant material and in fauna (for instance small shells) can be used to calculate historical mean CO_2 concentrations (Enoch et al., 1984).

TABLE 2. Isotopic composition of carbon dioxide. Data from Giggenbach (1990), Sigurdsson et al. (1987) and Kling et al. (1987).

	$\delta^{13}C$	^{14}C conventional date
MONOUN	$-7.2\ ^0/_{00}$ PDB	18,000 years BP
NYOS	$-3.4\ ^0/_{00}$ PDB	>35,000 years BP

3) CO_2 experimental enrichment on managed and unmanaged ecosystems

CO_2 released from the gas extraction pipes in the lakes will be blown through plastic ducts to a set of greenhouses with different CO_2 concentrations. From the greenhouses, the air will be blown by large fans to open-top chambers and/or through thinner perforated plastic sleeves (0.01 to 0.05 m_2 cross section) placed on the ground in the open. We will attempt to control the CO_2 in the greenhouses and in open top chambers at four concentrations: ambient bulk air (356 ppmv), ambient +200; ambient +400 and ambient +800 ppmv. If the air exchange rate is maintained at a rate of 1 exchange per minute in the greenhouses (volume 4000 m^3: 1000 m^2 and 4 metres high), we will need 830, 1660 and 3320 tons of CO_2 yearly, which makes a total flow-rate of CO_2 of 5810 t y^{-1}. This amount is equivalent to 4.6% of the CO_2 currently dissolved in lake Nyos, thus permitting an experiment lasting theoretically over 20 years.

Obviously enough, however, the prime motive for degassing lakes Nyos and Monoun is the mitigation of the risk of recurrent gas disasters in Cameroon. The span of the operation should thus be as short as possible, with a flow-rate compatible with a safe degassing: the safety criterion is to maintain the stratification of the lakes throughout the degassing operation, and hence not to disrupt the lake stability in a way that might trigger an uncontrollable limnic eruption. It has been suggested that the degassing should last no more than one year at Monoun, and five years at Nyos. If these time spans seem too short for biological research, one must keep in mind that Lake Nyos is continuously fed with magmatic CO_2 at a rate estimated between 10,000 $t.y^{-1}$ (Evans et al., 1993) and 40,000 $t.y^{-1}$ (Nojiri et al., 1993). For lake Monoun, the figure is about 1200 t y^{-1} (G.W. Kling & W.C. Evans, personal communication). In the future, this carbon dioxide could be extracted as a renewable resource for scientific purposes, in a continuous and controlled manner.

The energy of gas expansion in the pipes could be harnessed to produce electricity through small hydroelectric turbines and generators. More probably, electricity for the field experiments would be provided by a diesel driven generator or by solar panels. Should the generator system fail, there will be a fail-safe system that will open vents placed in the wall of the greenhouses, and hence facilitate ventilation until electricity is re-established. CO_2 enrichment will cease during electricity failure. The advantage of large greenhouses is that we could accommodate many plots representing agricultural and natural plants from the local ecosystems.

By using open-top chambers, we could study the reaction of the ecosystem in situ and accommodate some plots with larger plants, like trees: the height of an open-top chamber may be increased with time as the tree grows. According to the experience of open-top chambers users in the temperate zone, the micro-environment can be expected to be modified as follows: CO_2 can be kept at the target concentration of $\pm$ 50 ppmv, temperature increases up to 3°C in the summer during midday hours, solar radiation is reduced by $\pm$ 20%, wind velocity is also reduced while relative humidity is increased (Leadley & Drake, 1993).

We propose to modify the design of the commonly used open-top chamber so that we can use them as CO_2-assimilation chambers by restricting the outflow (Leadley & Drake, 1993). Further, we propose to use the degassed water, pumped up from the lakes by the expansion of CO_2, to cool the air of the chamber on a heat exchanger during periods of high solar radiation. Air humidity in the chambers may, at certain conditions, condense on the coolers and thus help to reduce air humidity. Plants should not be irrigated with CO_2 saturated irrigation water, as this will reduce their yield (Enoch & Olesen, 1993) but the upper (fresh water) layers of the lakes may meet irrigation needs.

Using this system, we expect to be able to maintain four replicate greenhouses and four replicate open-top chambers with three elevated CO_2 concentrations at 200, 400 and 800 ppmv above outside ambient CO_2 levels for a long period of time.

The experiments described here could be an important opportunity to learn about the ecological impact of elevated CO_2 in tropical ecosystems. The operation would present the advantage of eliminating a deadly risk for the local population and of performing at the same time a unique large scale and long term ecological experiment. It would also promote a North-South scientific cooperation: local scientists would work in CO_2 research together with experts of FACE experiments and international scientific teams.

REFERENCES

Bazzaz, F.A. (1990) The response of natural ecosystems to rising global CO_2 levels. *Annual Review of Ecology and Systematics*, **21**: 167-196.

Drake, B.G. (1992) A field study of the effects of elevated CO_2 on ecosystem processes in a Cheasapeake bay wetland. *Australian Journal of Botany*,. **40**: 579-595.

Enoch, Z.V., Carmi, I., Rounick, J.S.& Magaritz, M. (1984) Use of carbon isotopes to estimate incorporation of added CO_2 by greenhouse-grown tomato plants. *Plant Physiology*,. **76**: 1083-1085.

Enoch, H.Z. & Olesen, J.M. (1993) Plant response to irrigation with water enriched with carbon dioxide. *New Phytologist* Tansley review N° **54**, 125: 249-258.

Evans, W.C., Kling, G.W., Tuttle, M.L., Tanyileke, G. & White, L.D. (1993) Gas build up in Lake Nyos, Cameroon: the recharge process and its consequences. *Applied Geochemistry*, **8**: 207-221.

Giggenbach, W.F. (1990) Water and gas chemistry of Lake Nyos and its bearing on the eruptive process. *Journal of Volcanology and Geothermal Research,* **42**: 337-362.

Halbwachs, M., Grangeon, J., Sabroux, J.C. & Villevieille, A. (1993) Purge par auto-siphon du gaz carbonique dissous dans le lac Monoun (Cameroun): premiers résultats expérimentaux. *Comptes Rendus de l'Académie des Sciences.* Paris, **316/II**: 483-489.

Halbwachs, M., Sabroux, J.C., Vandemeulebrouck, J., Naah, E. & Villevieille, A. Continuous monitoring of the temperature, and field measurement of the CO_2 concentration profile in lakes Nyos and Monoun, Cameroon: preliminary steps for controlled degassing of the "Killer lakes". *In preparation.*

Kling, G.W., Clark, M.A., Compton, H.R., Devine, J.D., Evans, W.C., Humphrey, A.M., Koenigsberg, E.J., Lockwood, J.P., Tuttle, M.L. & Wagner, G.N. (1987) The 1986 Lake Nyos gas disaster in Cameroon, West Africa. *Science,* **236**: 169-175.

Kling, G.W., Evans, W.C. & Tuttle, M.L. (1991) A comparative view of Lakes Nyos and Monoun, Cameroon, West Africa. *Proceedings of the International Association for Theoretical and Applied Limnology*, **24**: 1102-1105.

Körner, C. & Arnone III, J.A. (1992) Responses to elevated carbon dioxide in artificial tropical ecosystems. *Science*, **257**: 1672-1675.

Leadley, P.W. & Drake, B.G. (1993) Open top chambers for exposing plant canopies to elevated CO_2 concentrations and for measuring net gas exchange. *Vegetatio.* **104**: 3-17.

Le Guern, F., Tazieff, H & Faivre-Pierret, H. (1982) An example of health hazard: people killed by gas during a phreatic eruption, Dieng Plateau, Java (Indonesia), February 20th, 1979. *Bulletin of Volcanology,* **45**: 153-156.

Le Touzey, R. (1985) Notice de la carte phytogéographique du Cameroun au 1:500,000. Institut de la Carte Internationale de la Végétation (Toulouse, France) et Herbier National (Yaoundé, Cameroun).

Nojiri, Y., Kusakabe, M., Tietze, K., Hirabayashi, J., Sato, H., Sano, Y., Shinohara, H., Njine, T. & Tanyileke, G. (1993) An estimate of the CO_2 flux in lake Nyos, Cameroon. *Limnology and Oceanography.* **38/4**: 725-739.

Oechel, W.C. & Billings, W.D. (1992) Effects of global change on the carbon balance of arctic plants and ecosystems. In *Arctic ecosystems in a changing climate* (eds. T. Chapin, R. Jeffries, J. Reynolds, G. Shower & J. Svoboda), pp. 139-168. Academic Press, New York.

Sabroux, J.C., Dubois, E. & Doyotte, C. (1987) The limnic eruption: a new geological hazard? Abstract, International Conference on Lake Nyos disaster, Yaoundé, Cameroon, March 16-20.

Scarascia Mugnozza, G., De Angelis, P., Matteucci, G., & Valentini, R. (1996)

Long term exposure to elevated CO_2 in a natural *Quercus ilex* L. community: Net photosynthesis and photochemical efficiency of PSII at different levels of water stress. *Plant, Cell and Environment*, in press.

Sigurdsson, H., Devine, J.D., Tchoua, F.M., Presser, T.S., Pringle, M.K.W. & Evans, W. (1987) Origin of the lethal gas burst from Lake Monoun, Cameroon. *Journal of Volcanology and Geothermal Research*, **31**: 1-16.

Stager, C. (1987) Silent death from Cameroon's Killer Lake. *National Geographic Magazine*, **172**: 404-420

Strain, B. & Cure, J.D. (1985) *Direct effect of increasing carbon dioxide on vegetation*. US Department of Energy, DOE/ER 0238, Washington DC.

Strain, B.R. & Thomas, R.B. (1992) Field measurements of CO_2 enhancement and climate change in natural vegetation. *Water Air and Soil Pollution*, **64**: 45-60.

Tietze, K. (1992) Cyclic gas bursts : are they a "usual" feature of Lake Nyos and other gas-bearing lakes? In *Natural hazards in West and Central Africa* (eds. S.J. Freeth, C.O. Ofoegbu & K. Mosto Onuoha), pp. 97-107. Vieweg, Braunschweig/ Wiesbaden.

Burning coal seams in southern Utah: a natural system for studies of plant responses to elevated CO_2

J. R. EHLERINGER, D. R. SANDQUIST AND S. L. PHILIPS

Stable Isotope Ratio Facility for Environmental Research, Department of Biology, University of Utah, Salt Lake City, Utah 84112, USA.

SUMMARY

In the Burning Hills and Smoky Mountains of southern Utah (USA), coal deposits exposed to the surface have been ignited by lightning and have been burning for periods of years to over a century. We examined one of these sites, where the below-ground combustion of this low-sulfur coal releases gases to the atmosphere from vents above the burning seam. The surrounding vegetation is cold-desert shrub, typical of the region and consisted of both C_3 and C_4 perennial species.

Additionally, at least one weedy C_4 species had invaded disturbed locations immediately adjacent to the active vent area. Atmospheric CO_2 concentrations in the vicinity of the vents fluctuated significantly, however, CO_2 concentrations measured approximately 500 m from the most active vents were 7 ppm elevated above ambient concentrations measured at a control site 10 km from the burning vents. CO_2 concentrations at sites nearer the vents, but still with natural, undisturbed vegetation, were elevated 65 ppm above ambient background values. At vegetated sites nearest the vents, CO_2 concentrations were elevated by an average of 542 ppm above ambient values.

The continuous distribution of C_4 vegetation along the CO_2 concentration gradient provides a means of estimating the long-term integrated CO_2 concentrations at each location. Using the carbon isotope ratio of the C_4 vegetation (*Atriplex confertifolia* and *Salsola iberica*) to estimate the atmospheric CO_2 concentration, we observed that the ratio of intercellular to atmospheric CO_2 concentrations of C_3 vegetation decreased in response to elevated CO_2 concentrations. This decreased ratio for *Gutierrezia sarothrae* (C_3) was sufficient to result in a predicted doubling of water-use efficiency.

INTRODUCTION

Doelling & Graham (1972) described extensive coal deposits in southern Utah, which underlie existing sandstone but which frequently outcrop at the surface. Occasionally the coal deposits at the surface are ignited by lightning, resulting in underground fires. In the Burning Hills and Smoky Mountains of southern Utah, a number of these fires have been burning for many years, releasing CO_2 and other gases into the atmosphere. The plant communities surrounding these coal vents are cold-desert vegetation, consisting of both C_3 and C_4 shrub and perennial herb species (Caldwell, 1985; West, 1988; Comstock & Ehleringer, 1992). Gleason & Kyser (1984) reported that the carbon isotope ratio ($\delta^{13}C$) of CO_2 released from one of the burning coal sites was -32.5 ‰, which is significantly more negative than atmospheric values which tend to be near -8 ‰. For the vent system they investigated, the CO_2 released from this combustion had the potential for impacting the $\delta^{13}C$ values of vegetation at distances of up to 700-800 m from the vents. Our objective was to quantify the CO_2 concentrations and evaluate through stable isotope analyses the potential of these vent sites as natural field experiments for studying the influence of elevated atmospheric CO_2 conditions on growth of desert vegetation.

Carbon isotope ratios in C_3 plants are influenced by the isotope ratio of the source air, fractionation steps between the bulk atmosphere CO_2 and the initial photosynthetic reaction by RuBP carboxylase (Rubisco), and by the drop in CO_2 concentration between the air and the intercellular spaces within the leaf. The factors influencing C_3 leaf carbon isotope ratios can be described as

$$\delta^{13}C_{plant} = \delta^{13}C_{air} - a - (b_3 - a)c_i/c_a, \qquad [1]$$

where $\delta^{13}C_{plant}$ is the carbon isotope ratio (‰ relative to the PDB standard) of the plant sample, $\delta^{13}C_{air}$ is the carbon isotope ratio of CO_2 in the atmosphere (typically -8 ‰ in a non-enriched atmosphere), a is the fractionation associated with the slower diffusion of $^{13}CO_2$ air (4.4 ‰), and b_3 is the net fractionation associated with Rubisco (27 ‰), and c_a and c_i are the atmospheric and intercellular CO_2 concentrations, respectively

(Farquhar, O'Leary & Berry, 1982). Together the terms ($-a - (b_3 - a)c_i/c_a$) are the discrimination (Δ) against ^{13}C by the leaf and represent the change in ^{13}C content between the atmosphere and fixed carbon in the leaf.

In C_4 plants, the initial carboxylation is by PEP carboxylase (PEPC), which discriminates much less than RuBP carboxylase (Rubisco), and the c_i/c_a ratio remains nearly constant because of high PEPC activity. Farquhar (1983) modeled the carbon isotope ratio of C_4 plants as

$$\delta^{13}C_{plant} = \delta^{13}C_{air} - a - (b_4 - b_3f - a)\, c_i/c_a \quad [2]$$

where b_4 is the net fractionation associated with PEP carboxylation (-5.7 ‰) and f is the bundle sheath leakiness, or the proportion of carbon fixed by PEPC that subsequently leaks out of the bundle sheath thereby allowing limited expression of Rubisco discrimination (b_3). The leakiness may also be thought of as a measure of "overcycling" by PEPC that occurs in the mesophyll cells, raising c_i in the bundle sheath cells (Farquhar, 1983). The sum ($b_4 - b_3f - a$) is a small number, further reduced by its product with the c_i/c_a ratio. As a consequence, $\delta^{13}C$ values of C_4 plants reflect the $\delta^{13}C$ of air with limited physiological influences.

Given this, it is likely that, at least within a species, the $\delta^{13}C_{plant}$ value of C_4 plants can be used to estimate $\delta^{13}C_{air}$ if we assume that other parameters in Eqn. 2 are constant. Evans et al. (1986) and Henderson, von Caemmerer & Farquhar (1992) provided initial evidence in support of this assumption. Marino & McElroy (1991) and Marino et al. (1992) have adopted this approach and used it to estimate $\delta^{13}C_{air}$ over historical time periods, although it is possible that some environmental influences could affect carbon isotope discrimination by C_4 plants (Henderson et al., 1992). There is evidence that soil salinity can influence carbon isotope discrimination by C_4 plants (Walker & Sinclair 1992; Sandquist & Ehleringer, 1994, Leffler, Basha & Ehleringer, 1996), but drought does not appear to have any impact (Leffler et al., 1996).

We examined the carbon isotope variation of C_4 vegetation on nonsaline sites at varying distances from one of the burning coal-seam vent sources to examine the possibility that atmospheric CO_2 concentrations could be reconstructed from C_4 isotopic values and that these vents might serve as a tool for investigating elevated CO_2 effects on ecophysiological activity of C_3 and C_4 species in this desert community. Atmospheric gases were also analyzed with an infrared gas analyzer to compare with isotopic observations.

MATERIALS AND METHODS

Study site

Field observations were made on an isolated, unnamed peak in the Burning Hills of southern Utah, USA (lat. 37° 16', long. 111° 22', 1750 m elevation). The coal seam on this peak had been burning for an extended period (>25 years) and was very near the site described by Gleason & Kyser (1984). We sampled vegetation at varying distances from a series of coal vents burning at the southern end of this peak; each sampling site was approximately 10 m in diameter.

Along the mountain there were occasional large crevices which emitted CO_2 at apparent lower rates than the large vents. The vegetation was cold desert perennial scrub and was dominated by *Agropyron desertorum* (C_3), *Artemisia tridentata* (C_3), *Atriplex confertifolia* (C_4), *Bromus tectorum* (C_3), *Chrysothamnus viscidiflorus* (C_3), *Coleogyne ramosissima* (C_3), *Elymus elymoides* (C_3), *Ephedra nevadensis* (C_3), *Eurotia lanata* (C_3), *Gutierrezia sarothrae* (C_3), *Juniperus osteosperma* (C_3), *Munroa squarrosa* (C_4), *Oryzopsis hymenoides* (C_3), *Opuntia* sp. (CAM), *Salsola iberica* (C_4), and *Sphaeralcea* sp. (C_3).

All sites sampled along this transect, except for the last one, consisted of undisturbed vegetation. The last site was nearest the active vents and had the highest CO_2 concentrations; there were obvious indications that this last site had been previously disturbed in the past by land-management attempts to bulldoze the area and extinguish the burning coal.

Atmospheric CO_2 measurements

Atmospheric CO_2 concentrations were measured with an infrared gas analyzer (LI-6200, Licor, Lincoln, NE, USA). At each site, CO_2 was measured every 1 s for 60 s and averaged to derive a single sample value. This sampling was repeated six to nine times per site to get an overall estimate of the mean CO_2 concentration and variability at a site. During the entire sampling period, the wind was moderate (1-3 m s^{-1}) from the south, which is typical of long-term patterns (Gleason & Kyser, 1984).

Carbon isotope composition of dry leaf matter

Leaf samples were taken from five stems on each of five plants per species at a site. The material from individual plants was bulked to form a single composite sample. All samples were oven dried at 85°C for forty-eight hours. The dried leaves were ground with liquid nitrogen to form a fine powder. Approximately 2 mg of powder was loaded onto an elemental analyzer coupled to an isotope ratio mass spectrometer (C. F. Kitty, delta S, Finnigan-MAT, San Jose, CA, USA). Samples were combusted, CO_2 was isolated, and then this gas was introduced into the mass spectrometer for carbon isotope ratio analysis. The carbon isotope ratio was calculated using "delta" notation as

$$\delta = [(R \text{ sample} / R \text{ standard})] \times 1000^0/_{00} , \qquad [3]$$

where R is the molar $^{13}C/^{12}C$ ratio of either the sample or standard (PDB). Units are parts per mil (‰). While the instrument has a repeatability of ± 0.02‰, the long-term, overall precision of sampling, combustion, and analysis is ± 0.11‰. The carbon isotope discrimination (Δ) of plant material (Farquhar, Ehleringer & Hubick, 1989) was calculated as

$$\Delta = (\delta^{13} C_{air} - \delta^{13} C_{plant}) / (1 + \delta^{13}C_{plant}), \qquad [4]$$

where $\delta^{13}C_{plant}$ was derived from $\delta^{13}C$ measurements as described above and $\delta^{13}C_{air}$ was derived as described below.

The δ_{air} value was not directly measured, but instead was calculated as a residual by rearranging the terms in Eqn. 2. Marino & McElroy (1991) had capitalized on the near constant photosynthetic fractionation by C_4 plants (all terms influencing $\delta^{13}C_{plant}$, except $\delta^{13}C_{air}$ in Eqn. 2 by measuring $\delta^{13}C_{plant}$ and assuming a constant photosynthetic fractionation; they calculated $\delta^{13}C_{air}$ as a residual. We calculated the $\delta^{13}C_{air}$ value in this manner. The baseline C_4 discrimination value was calculated using the $\delta^{13}C_{plant}$ value measured in a "clean" site 10 km away from the burning coal seam.

RESULTS AND DISCUSSION

There was no active vegetation in the area immediately adjacent to the most active burning coal vents and ambient atmospheric CO_2 concentrations decreased with distance from this location. The vegetation closest to the primary vents was approximately 50 m distant, and the atmospheric CO_2 concentration surrounding this vegetated site averaged 895 ppm. There was substantial variation about this mean CO_2 concentration associated with shifting wind. Overall the atmospheric CO_2 near this vegetation varied ± 134 ppm (standard error). At nearby undisturbed sites, the atmospheric CO_2 concentrations were lower than 420 ppm and exhibited substantially less variability. Overall, there was a strong positive correlation between the mean CO_2 concentration at a site and the total range of CO_2 concentrations measured with the IRGA ($r = 0.986$, $n = 7$, $P < 0.01$). At the control site 10 km to the north, the atmospheric CO_2 concentration was 352.1 ± 0.6 ppm.

Carbon isotope ratios of the C_4 vegetation varied substantially in association with changes in atmospheric CO_2 concentration (Fig. 1). *Atriplex confertifolia* and *Salsola iberica* were the most common C_4 species along the gradient and exhibited $\delta^{13}C$ values ranging from -20.7 to -12.6 ‰. *Munroa squarrosa* occurred at three of the microsites sampled and its isotopic composition varied from -15.2 to -13.7 ‰ in a manner consistent with the isotopic changes in the other two C_4 species.

In order to calculate atmospheric CO$_2$ concentrations, the C$_4$ discrimination value was estimated from Eqn. 2 using the δ^{13}C values of C$_4$ plants from a control site 10 km away.

The δ^{13}C$_{air}$ for each site near the vent was then calculated by subtracting the C$_4$ discrimination value from the observed δ^{13}C$_{plant}$ values of the C$_4$ plants within these sites.

The atmospheric CO$_2$ concentrations were then calculated from a linear mixing model using the δ^{13}C$_{air}$, based on C$_4$ plants, and the bulk atmospheric air based on observations at the control site. The calculated CO$_2$ concentration compared favorably with IRGA observations at the different sites ($r = 0.797$, $P < 0.05$, Fig. 2).

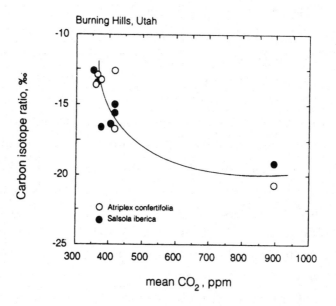

FIG. 1. The carbon isotope ratio of leaves of two C$_4$ species at different sites near a burning coal vent as a function of the atmospheric CO$_2$ as measured with an IRGA.

At slightly elevated atmospheric CO$_2$ concentrations (365-425 ppm), there was very close agreement between predicted and actual CO$_2$ concentrations. However, at the highest atmospheric concentrations, there was some discrepancy among methods, possibly as a result of turbulence-

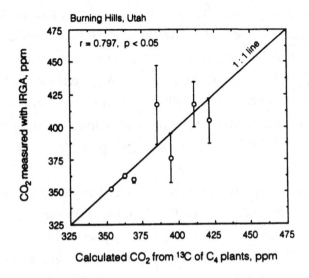

FIG. 2. The correlation between the measured atmospheric CO_2 (IRGA measurements) and the atmospheric CO_2 concentration calculated from the ^{13}C of C_4 plants at this site. Vertical lines indicate the 95 % confidence interval surrounding the IRGA CO_2 measurements.

based variations in the IRGA-measured atmospheric values. On the other hand, a second possibility is that carbon isotope discrimination by C_4 plants had increased at the site with most elevated CO_2 levels as predicted by Farquhar (1983), indicating increased expression of Rubisco discrimination.

Overall the CO_2 concentrations at these locations nearby the burning coal seam were significantly elevated above natural conditions, suggesting that sites in this area might be useful locations for evaluating plant performance under elevated CO_2 levels.

One possible limitation to this system could be complications arising from oxidation of sulfur within the coal seams if baseline sulfur levels were high enough to cause SO_2-induced stomatal closure (Taylor & Pitelka, 1991). Atmospheric SO_2 levels were not directly measured, but from observations in the area, there were no indications of sulfur deposition near the soil surface. The coal seams contained less than 0.8% sulfur and are known to be nearly water saturated (Doelling & Graham, 1972).

It is possible that SO_2 formed during combustion processes was partly dissolved in deep soil water before these gases could reach the surface, thereby minimizing SO_2 effects. The Burning Hills of southern Utah are distinct from similar burning coal seams in other parts of the world (e.g., arctic Smoking Hills as described by Havas & Hutchinson, 1983) in that the Utah coal seams are substantially lower in sulfur content. While it is not possible to eliminate SO_2 effects from confounding data interpretation, the available evidence suggested that this possible impact would be minimal.

Gutierrezia sarothrae was the only C_3 species distributed continuously along this CO_2 gradient. Leaf carbon isotope discrimination values (Δ) decreased from 20.2 to 13.5 ‰ as atmospheric CO_2 concentrations increased from 352 to 422 ppm (Fig. 3). The changes in the calculated leaf δ values were equivalent to a decrease in the c_i/c_a ratio from 0.70 to 0.40 (Eqn. 1). This change in c_i/c_a ratio was large enough that it should have significant impact on plant gas exchange rates (Farquhar *et al.*, 1989) as well as resulting in a doubling of plant water use efficiency in plants along this gradient (Ehleringer *et al.*, 1992).

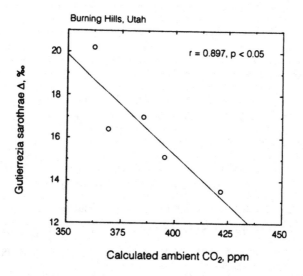

FIG. 3. The relationship between carbon isotope discrimination in *Gutierrezia sarothrae* (a C_3 plant) and atmospheric CO_2 concentration as estimated by the ^{13}C content of surrounding C_4 species at the site.

At the community level, carbon isotope discrimination values decreased as atmospheric CO_2 levels increased. For plants still active at the August sampling period (*Agropyron, Coleogyne, Ephedra, Gutierrezia,* and *Oryzopsis*), leaf δ values were significantly correlated with the CO_2 concentrations at the respective sites. The correlation was highly significant ($r = 0.576$, $P < 0.02$, $n = 18$) despite large differences in the individual values among species.

A reduction in stomatal conductance is typical in leaves of C_3 plants exposed to elevated CO_2 (Cure & Acock, 1986; Bunce, 1993). Such a mechanism could easily account for the observed decreases in the c_i/c_a ratio of *G. sarothrae*. However, some caution may be necessary when interpreting these initial data. These burning coal vents are characterized by an incomplete combustion, resulting in a release of saturated hydrocarbons in addition to CO_2. Using measured vent concentrations from Gleason & Kyser (1984), we calculated that methane and saturated hydrocarbon concentrations at our sampled sites were in the neighborhood of 5 ppm. While we are unaware of any studies indicating that methane and other hydrocarbons at these concentrations have an adverse effect on stomatal activity, direct experimentation is necessary to test the possibility.

The significant decreases in the c_i/c_a ratio among all species as atmospheric CO_2 increased suggests that water-use efficiency will also have increased in these plants. Since soil moisture is the primary climatic factor limiting productivity in aridland ecosystems, any mechanism that decreases the rate of soil moisture extraction might provide a growth advantage to plants, most likely perennials, capable of utilizing the remaining moisture later in the season. While we have no data available yet showing that soil moisture is retained longer in this aridland ecosystem, Field *et al.* (1992) have suggested that in semiarid ecosystems stomatal closure associated with elevated CO_2 should result in greater soil moisture retention at depths not subject to surface evaporation and thus possibly extend productivity to later periods in the drought.

The burning coal vents of southern Utah have the potential to serve as a vehicle for examining elevated atmospheric CO_2 effects on aridland vegetation. Given the low productivity and low cover that typically characterize aridland ecosystems, these communities are not likely to be a high priority agenda item for large-scale elevated CO_2 manipulations, such as free-air-CO_2-enrichment (FACE) studies. Yet aridland ecosystems may be particularly responsive to elevated atmospheric CO_2 if stomatal closure results in a reduced rate of soil moisture loss. Whereas in some natural burning-coal vents, high-sulfur coals result in SO_2-pollution effects that compound interpretation of plant responses (Freedman *et al.*, 1990), this may not be the case for the coal seams of southern Utah.

In the remote Burning Hills of southern Utah, lightning strikes have resulted in several burning seams. Doelling & Graham (1972) reported numerous sites where the coal seam had ignited and later burned out, indicating that this process has been ongoing for extended time periods. More than four large, long-term burns are known to occur at present and other small, unknown burns may exist as well. Of the large burns, at least one is thought to be more than a century old.

Burning coal vents are somewhat analogous to the elevated-CO_2 springs described by Miglietta *et al.*, (1993). These geothermal sources occur throughout the Mediterranean-climate landscape in central Italy and provide a resource for evaluating plant acclimation and adaptation to elevated atmospheric CO_2 conditions. In both systems, atmospheric turbulence will result in fluctuations in the absolute CO_2. While the extent to which stomata can respond to high frequency changes in CO_2 concentration is unknown, carbon isotope ratio values of C_4 plants will provide an estimate of the long-term, assimilation-weighted value of the atmospheric CO_2 concentration.

ACKNOWLEDGEMENTS

This research has been supported by the Ecological Research Division of the Office of Health and Environmental Research at the U.S. Department of Energy.

REFERENCES

Bunce, J.A. (1993) Effects of doubled atmospheric carbon dioxide concentration on the responses of assimilation and conductance to humidity. *Plant, Cell and Environment*, **16**, 189-197.

Caldwell, M.M. (1985) Cold desert, In *Physiological Ecology of North American Plant Communities* (eds. B.F. Chabot & H.A. Mooney), pp. 198-212. Chapman and Hall, London.

Comstock, J.P. & Ehleringer, J.R. (1992) Plant adaptation in the Great Basin and Colorado Plateau. *Great Basin Naturalist*, **52**, 195-215.

Cure, J.D. & Acock, B. (1986) Crop responses to carbon dioxide doubling: a literature survey. *Agricultural and Forest Meteorology*, **38**, 127-145.

Doelling, H.H. & Graham, R.L. (1972) Southwestern Utah Coal Fields: Alton, Kaiparowits Plateau and Kolob-Harmony. *Utah Geological and Mineralological Survey*. Monograph Series 1, Salt Lake City. 33 pp.

Ehleringer, J.R., Phillips, S.L. & Comstock, J.P. (1992) Seasonal variation in the carbon isotope composition of desert plants. *Functional Ecology*, **6**, 396-404.

Evans, J.R., Sharkey, T.D., Berry, J.A. & Farquhar, G.D. (1986) Carbon isotope discrimination measured concurrently with gas exchange to investigate CO_2 diffusion in leaves of higher plants. *Australian Journal of Plant Physiology*, **13**, 281-292.

Farquhar, G.D. (1983) On the nature of carbon isotope discrimination in C_4 species. *Australian Journal of Plant Physiology*, **10**, 205-226.

Farquhar, G.D., Ehleringer, J.R. & Hubick, K.T. (1989) Carbon isotope discrimination and photosynthesis. *Annual Review of Plant Physiology and Molecular Biology*, **40**, 503-537.

Farquhar, G.D., O'Leary, M.H. & Berry, J.A. (1982) On the relationship between carbon isotope discrimination and intercellular carbon dioxide concentration in leaves. *Australian Journal of Plant Physiology*, **9**, 121-137.

Field, C.B., Chapin, F.S. III, Matson, P.A. & Mooney, H.A. (1992) Responses of terrestrial ecosystems to the changing biosphere: a resource-based approach. *Annual Review of Ecology and Systematics*, **23**, 201-235.

Freedman, B., Zobens, V., Hutchinson, T.C. & Gizyn, W.I. (1990) Intense, natural pollution affects arctic tundra vegetation at the Smoking Hills, Canada. *Ecology*, **71**, 429-503.

Gleason, J.D. & Kyser, T.K. (1984) Stable isotope compositions of gases and vegetation near naturally burning coal. *Nature*, **307**, 254-257.

Havas, M. & Hutchinson, T.C. (1983) The Smoking Hills: natural acidification of an aquatic ecosystem. *Nature, **30**, 23-27.

Henderson, S.A., von Caemmerer, S. & Farquhar, G.D. (1992) Short-term measurements of carbon isotope discrimination in several C_4 species. *Australian Journal of Plant Physiology*, **19**, 263-285.

Leffler, A.J., Basha, F. & Ehleringer, J.R. (1996) Salinity increases bundle sheath leakiness in *Atriplex canescens*, a C_4 halophyte. *Plant, Cell and Environment* (in review).

Marino, B.D. & McElroy, M.B. (1991) Isotopic composition of atmospheric CO_2 inferred from carbon in C4 plant cellulose. *Nature*, **349**, 127-131.

Marino, B.D., McElroy, M.B., Salawitch, R.J. & Spaulding, E.W.G. (1992) Glacial-to-interglacial variations in the carbon isotope composition of atmospheric CO_2. *Nature*, **357**, 461-466.

Miglietta, F., Raschi, A., Bettarini, I., Resti, R. & Selvi, F. (1993) Natural CO_2 springs in Italy: a resource for examining long-term response of vegetation to rising atmospheric CO_2 concentrations. *Plant, Cell and Environment*, **16**, 873-878.

Sandquist, D.R.& Ehleringer, J.R. (1994) Carbon isotope discrimination in the C_4 shrub *Atriplex confertifolia* (Torr. & Frem.) Wats. along a salinity gradient. *Great Basin Naturalist*, **5**, 135-141.

Taylor, G.E.& Pitelka, L.F. (1991) *Ecological Genetics and Air Pollution*. Springer Verlag, New York.

Walker, C.D. & Sinclair, R. (1992) Soil salinity is correlated with a decline in ^{13}C discrimination in leaves of *Atriplex* species. *Australian Journal of Ecology*, **17**, 83-88.

West, N.E. (1988) Intermountain deserts, shrub steppes, and woodlands. In *North American Terrestrial Vegetation* (eds. M.G. Barbour & W.D. Billings), pp. 283-306. Cambridge University Press, New York.

Long-term effects of enhanced CO_2 concentrations on leaf gas exchange: research opportunities using CO_2 springs

P.R. van GARDINGEN[1], J. GRACE[1], C.E. JEFFREE[2], S.H. BYARI[3], F. MIGLIETTA[4], A. RASCHI[4] and I. BETTARINI[4]

[1] *Institute of Ecology and Resource Management,*
University of Edinburgh, UK.
[2] *Science Faculty, Electron Microscopy Unit,*
University of Edinburgh, UK.
[3] *Faculty of Meteorology, Environment and Arid-land Agriculture,*
King Abdulazziz University, Jeddah, Saudi Arabia.
[4] *Institute of Environmental Analysis and Remote*
Sensing for Agriculture, CNR-IATA, Firenze, Italy.

SUMMARY

The Bossoleto site in Tuscany, central Italy has been the location of a two year research programme investigating the long-term effects of enhanced CO_2 concentrations on natural vegetation. Research has investigated the effects of CO_2 on leaf gas exchange on a grass *Phragmites australis* (Cav.) Trin. ex Steudel and the tree species *Quercus pubescens* Willd. Preliminary results for *P. australis* demonstrate reductions in stomatal density, stomatal conductance, and maximum photosynthetic rates when compared with nearby control sites. In contrast, work with *Q. pubescens* showed no evidence of photosynthetic acclimation. The responses observed in plants growing at the Bossoleto site are consistent with results from short-term experiments, providing some confidence in this approach. The characteristics of CO_2 springs can add to the challenge of designing effective experiments. It is difficult to locate control sites that have similar vegetation, soil type and environmental characteristics. When working with natural vegetation, variability in characteristics within and between populations, can exceed any response to CO_2. Atmospheric concentrations can vary at time scales ranging from the order of seconds through to seasons. High CO$_2$ concentrations at some sites may be associated with pollutants that can affect plant growth. These potential problems influence the possible types of experimental approaches. This paper uses preliminary analysis of leaf gas exchange data from the Bossoleto site, to examine how natural sources of CO$_2$ enrichment can be used in biological research.

INTRODUCTION

How can natural sources of CO$_2$ be used in biological research? The likely effects of enhanced atmospheric CO$_2$ concentrations on vegetation have been the subject of intensive research for over a decade. Work has described the short-term effects of CO$_2$ on many physiological processes of plants, with marked emphasis on growth, water use and carbon assimilation and partitioning. The long-term effects of CO$_2$ on vegetation has remained a relatively under-researched topic, mainly because of technical difficulties and the high cost of experimentation.

Field experiments investigating effects of enhanced CO$_2$ concentrations on plants and animals tend to be technically challenging (Lawlor & Mitchell, 1991; Allen *et al.*, 1992; Leadley & Drake, 1993) and expensive (Kimball, 1992), leading to very few long-term or large-scale studies. Natural CO$_2$ vents offer an alternative method to study the effects of long-term exposure on ecosystems (Miglietta *et al.*, 1993a; Miglietta *et al.*, 1995; Miglietta *et al.*, 1993b; Miglietta & Raschi, 1993). It should be stressed that an investigation of ecological processes in these systems can be justified on scientific grounds without reference to the obvious link to the issue of global environmental change. It is obvious, however, that this is the reason for increasing research activity at these sites.

One attraction of these sites, has been the knowledge that vegetation has been exposed to high CO$_2$ for long periods, allowing time for acclimation, and perhaps genetic adaptation. The main disadvantage is the difficulty in defining experimental procedures. CO$_2$ concentrations fluctuate over various time-scales, and with distance from the source and may be confounded with other environmental factors such as soil type or presence of saline ground water at mineral springs. As a result, it can be difficult to find suitable control sites which have similar environmental and biological characteristics.

The purpose of this chapter, is to consider how the environmental characteristics of CO_2 springs are likely to influence plant growth, with emphasis on leaf physiology. The results from a two year study at the field site, Il Bossoleto in Tuscany, central Italy, are used to demonstrate how CO_2 springs can be used and to illustrate some difficulties associated with this approach. The chapter concludes with guidelines for designing effective experiments, and suggestions for future research applications.

CHARACTERISTICS OF THE SITE

Ecosystems developing near CO_2 springs are affected by environmental gradients often leading to distinct zonation of vegetation. Several CO_2 springs in central Italy, including the Bossoleto site used in the current study have been described previously (Miglietta *et al.*, 1993b). The Bossoleto site (Lat. 43° 17', Long. 11° 35') is located approximately 40 km south-east from the city of Siena. It is situated in a small bowl-like depression of approximately 80 m diameter and 20 m depth (Fig. 1). The site has three main zones of vegetation (Miglietta *et al.*, 1993b). At the bottom there are extensive areas of *Phragmites australis* growing in dense mono-specific stands. Around the top of the bowl is an area of mixed woodland, dominated by *Quercus ilex* L. and *Q. pubescens* Willd. Lastly there is a small area of calcareous grassland (marked Gr in Fig. 1).

A number of CO_2 vents are distributed around the Bossoleto site, but emissions are dominated by a large vent in a cave at the bottom of the bowl, towards the south of the site (Fig. 1). In addition there are two mineral springs at the bottom of the bowl. The CO_2 from the vents is over 99 % pure and has low levels of pollutants such as H_2S and SO_2 often associated with geothermal sources. Concentrations of these gases have been monitored using diffusion tubes (Dräger Aktiengesellschaft, Germany) and found to be virtually undetectable.

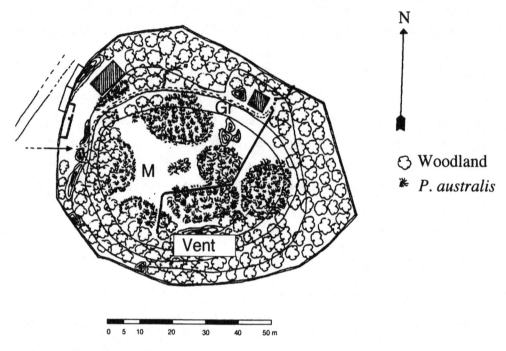

FIG. 1. Il Bossoleto. Main vegetation types are shown. The small grassland community is marked Gr. The main CO$_2$ vent is in a cave under the position marked Vent. The position marked M shows the location of the mast used for measuring the CO$_2$ profile. Contour lines represent 4 m height.

The water table is close to the surface of the soil and plant roots are in   contact with a supersaturated salt solution at 38 °C. The ionic composition of the water mainly comprises of the cations Ca^{2+}, Mg^{2+}, Na$^+$ and the anions HCO$_3^-$, SO$_4^{2-}$ and Cl$^-$. The solid residue was over 5 g l^{-1}. Atmospheric CO$_2$ concentrations vary during the day and partly depend on atmospheric conditions. Concentrations as high as 75 % have been measured between 7 and 8 am (van Gardingen *et al.*, 1995). At this time of day, it was possible to see the refractive boundary between the region with very high CO$_2$ concentrations at the bottom of the bowl and more normal, less dense, air above it. The boundary shimmered as a result of the optical densities of the two air masses, and mist often floated at the interface. The boundary tended to become unstable as soon as direct solar radiation was incident on the zone at the bottom of the bowl at around 9 am. The high concentrations then rapidly dispersed dropping to levels of around 2000 μmol mol[-1] at the centre of the bowl after around one or two hours. A CO_2 profile is shown for two sampling periods on July 31 1992 in Fig. 2. The zone at the bottom of the bowl is dominated by monospecific stands of *P. australis* growing on peat and undecayed litter. The upper boundary of this zone has been observed to coincide with the top of the dense pool of high CO_2 concentrations in the early morning. It is interesting to speculate that this feature of the environment may explain why *P. australis* grows alone at the Bossoleto site, while elsewhere in the region it forms part of more species-rich wetland communities. Certainly, it is known that *P. australis* has the ability to tolerate exposure to anaerobic and saline conditions (Matoh, Matsushita & Takahashi, 1988; Crawford, 1989; Hellings & Gallagher, 1992; Weisner, Graneli & Ekstam, 1993). The presence of "indicator species" such as *Agrostis canina* L. growing near CO_2 vents has been reported previously (Miglietta *et al.*, 1993b). In Italy, this species is usually found only at these sites, indicating that the modified environment is more favourable for survival. The challenge in ecological studies of CO_2 springs is to determine which of several co-varying environmental factors control distribution of vegetation.

Other environmental factors at the Bossoleto site are influenced by the combination of topography and CO_2 emissions.

The most marked example is observed for air temperature which shows a rapid increase in the early morning associated with the presence of the stable atmospheric conditions at the bottom of the bowl (Fig. 3). This heating is associated with an enhanced greenhouse effect, resulting from the high CO_2 concentrations at the bottom of the bowl.

The stability decreases only after air temperature has increased sufficiently for buoyancy to promote mixing. As a result of this physical process, there is a period of over two hours each day when plants at the Bossoleto site experience the combination of high temperature, CO_2 concentration and humidity. Temperatures tend also to be higher in the afternoon because of the reduced wind speeds at the bottom of the bowl. It is difficult to assess how the combination of these factors are likely to

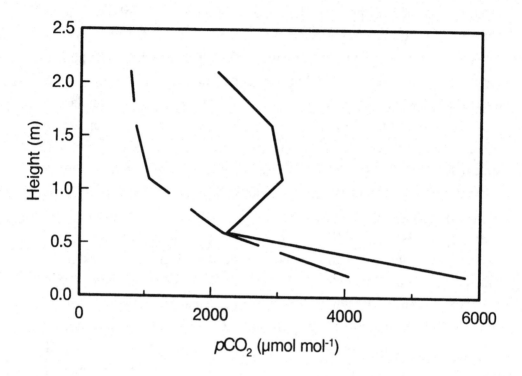

FIG. 2. Carbon dioxide profile measurements at Bossoleto during the day of July 31, 1992. (Solid line 11:30, dashed line 16:30).

affect plants. The increased temperature alone will affect the accumulated thermal time for plants and would be expected to influence events such as timing of flowering.

These examples have been used to illustrate the complexity of ecosystems surrounding CO$_2$ springs.

It is not advisable to assume that the vegetation pattern can be explained with sole reference to the gradient of average CO$_2$ concentration.

Instead, it is necessary to examine the concentration fluctuations of CO2 together with other environmental factors that co-vary over space and time.

This challenge for ecological science, creates even more difficulty for physiological studies since it is hard to define the growth conditions for plants when comparing between sites.

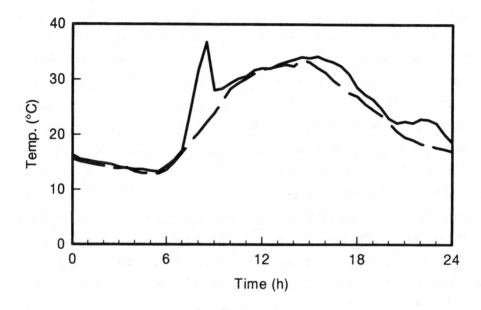

F IG. 3. Typical daily time course of air temperature at the Bossoleto (solid line) and Phragmites control (dashed line) sites. Temperatures were measured with fine unshielded thermocouples at a height of 1 m on separate days (Bossoleto, July 23, Phragmites control, July 25). Data were collected at location A at the Bossoleto site (Fig. 1) and are thirty minute means.

PHYSIOLOGICAL CHARACTERISTICS OF THE VEGETATION

The effects of CO_2 on vegetation have been studied extensively and reviewed by numerous authors (Acock, 1990; Bazzaz, 1990; Melillo *et al.*, 1990; Patterson & Flint, 1990; Woodward, 1992; Drake, 1992; Bowes, 1993; Kimball *et al.*, 1993). As a result there is a considerable body of data describing the responses of plants to enhanced CO_2 concentrations. Examination of the results can be confusing, since many appear to conflict, though part of the reason may be experimental artefacts such as the effects of pot size (Arp, 1991). There is also the ever present question, of how do results from short-term experiments relate to likely responses of ecosystems over periods of decades? How then, can natural CO_2 springs be used to improve our knowledge of physiological responses to CO_2? Vegetation growing around natural CO_2 springs has been exposed to enhanced CO_2 concentrations for periods much longer than possible in any experiment. While it may be difficult to define the average environmental conditions experienced by the plants, it is possible to compare physiological characteristics with plants growing at similar "control" sites. Differences between these two sets of plants, can then be compared with the types of responses observed in short-term, but more controlled experiments. This is the experimental approach that has been emphasised in the research at the Bossoleto site to date. Work has used measurements of leaf gas exchange, examination of leaf surface characteristics, and estimation of water use using an energy balance approach. In this section we present a preliminary analysis of these data, in order to illustrate the opportunities and difficulties in using CO_2 springs for physiological research. Leaf gas exchange was monitored using up to three LI-COR LI-6200 portable photosynthesis systems in 1992 and 1993. Measurements under ambient conditions were recorded at the Bossoleto and control sites in 1992 for *P. australis* and *Q. pubescens*. The systems were operated in closed mode a ¼ l chamber and results calculated using standard formulae (von Caemmerer & Farquhar, 1981). Data for *P. australis* collected in the afternoon of July 21 1992 at location A at the Bossoleto site (Fig. 1) are presented in Fig. 4. Ambient CO_2 concentrations fluctuated between 500 and >2000 µmol mol^{-1} (Fig. 4a). These data were obtained during leaf gas exchange measurements, but it was also possible to monitor CO_2 concentrations between measurements with an open chamber. Carbon dioxide concentrations fluctuated rapidly, often over periods of seconds, and occasionally exceeded 2000 µmol mol^{-1}, the upper range of the analyser's scale. The effects of the fluctuations in CO_2 concentration are seen in the leaf gas exchange measurements. Leaf photosynthesis (Fig. 4b) varied markedly, but not always as would be expected with CO_2 concentration (Fig. 4e).

Stomatal conductance was also variable (Fig. 4d), together leading to considerable scatter in the calculated values of C_i. Attempts to explain the variation in A and g_s in terms of CO_2 concentrations (Fig. 4e,f)

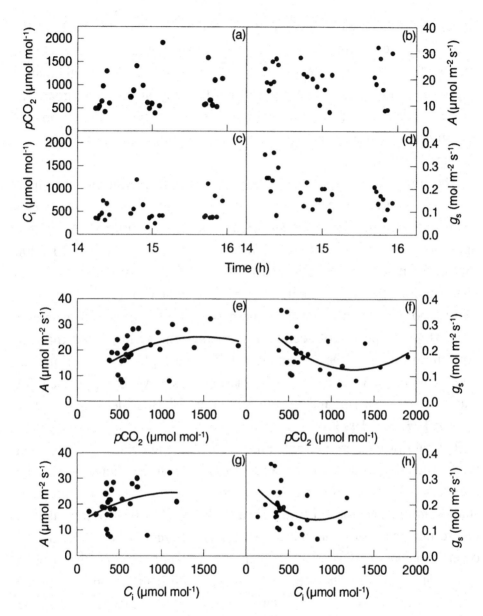

FIG. 4. Gas exchange data for leaves of *P. australis* growing at the centre of the Bossoleto site (Location A, Fig. 1). Data were obtained under ambient conditions, using a LI-COR LI-6200 portable photosynthesis system. Solid lines are second order polynomials fitted to the data. (a) Ambient atmospheric CO_2 concentration (pCO_2), (b) Leaf photosynthesis (A), (c) Leaf internal CO_2 concentration (C_i), (d) Stomatal conductance (g_s), (e) Relationship between A and pCO_2, (f) Relationship between g_s and pCO_2, (g) Relationship between A and C_i, (h) Relationship between gs and C_i.

or C_i (Fig. 4g,h) show some pattern, but again serve to illustrate the variability in the data. The relationship between A and C_i (Fig. 4g) is consistent with that shown in Fig. 5a obtained using an alternative approach. It is probable that part of the difficulty in obtaining good gas exchange data is the difference in time-scales between the period of measurement, fluctuations in ambient CO_2 concentrations and the responses of the leaves.

The difficulties associated with gas exchange measurements under ambient conditions necessitates alternative approaches. One method is to make measurements under steady state conditions. A related approach can produce A/C_i relationships in the field using the LI-6200 in closed mode (McDermitt et al., 1989). Leaves are placed in the chamber for at least 15 minutes with a constant CO_2 concentration of 1800 μmol mol^{-1} and a leaf-air vapour pressure difference of around 1 kPa. After this period of acclimation, the system is closed and the CO_2 concentration in the chamber falls through the action of photosynthesis. A series of A/C_i relationships have been derived for leaves of P. australis and Q. pubescens at the Bossoleto and control sites in 1992 and 1993.

Typical raw data are presented in Fig. 5, showing evidence of apparent photosynthetic acclimation to CO_2 for P. australis but not for Q. pubescens. These contrasting results are consistent with many previous studies, in that acclimation is not always observed. It is also important to realise that the results observed for P. australis could be influenced by other variables, particularly the high salt concentration of the ground water, and periodic anaerobic events. It is difficult to obtain reliable estimates of stomatal conductance using gas exchange under the conditions prevailing at the Bossoleto site. Data obtained under ambient conditions, were not reliable because of the CO_2 fluctuations. Data obtained during the A/C_i measurements, are considered unrealistic because of the rate of change of CO_2 concentration during the measurement period.

It is possible to estimate stomatal conductance using an energy balance approach from measurements of the temperature difference between transpiring and non-transpiring leaves (Leuning & Foster, 1990). This approach was used in 1992 to estimate stomatal conductance of leaves of

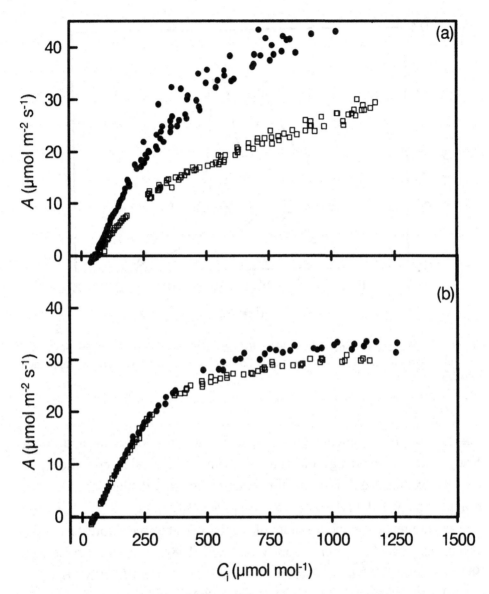

FIG. 5. Relationship between carbon assimilation (A) and calculated leaf internal CO_2 concentration (C_i) at the Bossoleto (□) and control (●) sites. (a) Leaves of *P. australis* (July 1992) (b) Leaves of *Q. pubescens* (May 1993).

P. australis. Typical data are shown in Fig. 6, demonstrating much lower conductances at the Bossoleto site.

It is however not clear, if the lower conductance results from the short-term response of stomata to CO_2, or a long-term response of leaf

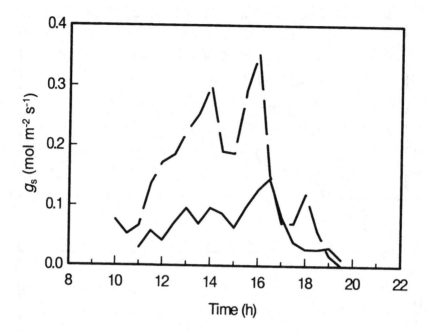

FIG. 6. Stomatal conductance of *P. australis* leaves measured at the Bossoleto (solid line, July 23) and control (dashed line, July 25) sites using an energy balance approach (Leuning & Foster, 1990).

growth and development. This question can only be answered using steady-state gas exchange measurements under controlled conditions, or from the examination of stomatal characteristics to calculate conductance from dimensional data (Parlange & Waggoner, 1970).

There are reports in the literature of relationships between stomatal density and atmospheric CO_2 concentration (Woodward, 1987; Woodward & Bazzaz, 1988; Peñuelas & Matamala, 1990; Rowland-Bamford *et al.*, 1990; Beerling & Chaloner, 1993b; Beerling & Chaloner, 1993a), and other reports of no apparent response to CO_2 (Radoglou & Jarvis, 1990). The stomatal density and stomatal index of *Q. ilex* growing near CO_2 springs have previously been reported (Miglietta *et al.*, 1993a) showing no difference from that observed at control sites. Stomatal density for *P. australis* (Table 1) was recorded using leaf surface replicas obtained using dental impression material (Wynn, 1976). A reduction in stomatal density was observed on the adaxial surface of plants growing at the Bossoleto site.

TABLE 1. Stomatal density (mm^{-2}) for *Phragmites australis* measured using leaf surface replicas. Data are the mean of 5 leaves (6 at the control site) ± 1 s.e. The locations at the Bossoleto site are shown on Fig. 1.

Location	Abaxial Surface	Adaxial Surface
Bossoleto, A	738±36	537±40
Bossoleto, B	921±52	713±50
Control	774±55	978±42

It is interesting to note that Woodward & Bazzaz (1988) report that the stomatal density on the adaxial surface of leaves from a range of species responded more to CO_2 concentration than the abaxial surface.

RESEARCH OPPORTUNITIES

Field based physiological research at sites with natural CO_2 enhancement will always be challenging because of the necessary compromises in experimental design. It is however possible to design simple experiments that compare physiological characteristics of plants growing near CO_2 springs with results from short-term controlled experiments. The results described above, demonstrate that CO_2 springs have a role in global environmental change research.

In the following sections, we discuss opportunities for future research, and ways of designing experiments that maximise the benefit of research on sites with natural CO enhancement.

This review has emphasised the use of natural CO_2 springs for studies on long-term effects of enhanced CO_2 concentrations on physiological characteristics of plants. This needs to be expanded to include other aspects of the system including soils, insects and to a lesser extent vertebrates.

Some of this work will be descriptive, but most should concentrate on investigating processes, for example, competition between plants for resources, or microbial decomposition of plant material. Process-based research can more easily be compared with results from other experimental approaches, and extrapolated to other sites or conditions. Natural CO_2 springs have considerable potential for studies of genetic adaptation to enhanced CO_2 concentrations. This application should be actively pursued, since CO_2 springs are apparently the only way that genetic studies will be possible over many generations for most species of plants.

Potential areas for research can be grouped into studies of the characteristics of natural ecosystems near CO_2 springs, and research using these systems as a tool to investigate global environmental change.

This distinction is not intended to suggest that ecological research on these systems cannot add to our understanding of likely global change. Instead it is meant to emphasise that the research objectives and approach are likely to differ.

EXPERIMENTAL DESIGN

The difficulties inherent in working with natural CO_2 springs can exacerbate the problem of experimental design. How can experiments be designed to make the best use of these natural resources? Clear objectives need to be linked to simple hypotheses so that effective experiments can be implemented.

There are two main problems that need to be overcome when designing experiments using natural CO_2 springs. Firstly, it is extremely difficult to define the environmental conditions at any location. This is linked to the second problem, in that several environmental factors co-vary in space and time, making it difficult to separate their effects. Good experimental design can minimise these problems.

The only way to define the environment, is to implement a comprehensive measurement programme. This needs to include standard parameters such as solar radiation, temperature and humidity, but also factors that will vary with distance from the source of CO_2. The environmental measurements will help when it is necessary to select control sites for experiments.

The atmospheric concentration of CO_2 needs to be measured and integrated over different time scales at each location. Conventional gas analysis techniques are suitable for short-term measurements, but others such as diffusion tubes or isotopic analysis of plant material (van Gardingen *et al.,* 1995) are required for integration over longer periods. Combined with these measurements, there should be analysis of atmospheric composition with emphasis on detecting potentially harmful pollutants. Chemical analysis of soil and plant material will provide additional information on nutritional status. Lastly a full botanical description should be prepared for each site. A comprehensive set of environmental data are required when interpreting results from experiments conducted at CO_2 springs. In the absence of such data, it is imprudent to ascribe observed differences to single factors such as CO_2 exposure.

The problems associated with co-variance of environmental factors are more difficult to solve. At sites such as Bossoleto the main co-varying factors are soil, water and CO_2. Environmental measurements and analysis of plant material can be used to describe the gradient and effects on plants, but it is impossible to remove the gradient from *in situ* experiments. This raises further ways that natural CO_2 springs can be used. Firstly, they can be used as an inexpensive source of CO_2 enrichment in field experiments.

Another application is to use the sites as a source of biological material for transfer to experiments under controlled conditions (Miglietta, *et al.,* 1992). Lastly it is possible to investigate genetic differences between populations in reciprocal transplant experiments.

CONCLUSIONS

The opportunities for research at natural CO_2 springs seem at first to be almost limitless, but the enthusiasm of potential researchers should be tempered by the realisation that it is difficult to design experiments that will achieve desired objectives.

The sites are a natural resource for global environmental change research, and can justify ecological research in their own right. Sites such as the Bossoleto, have complex relationships between several environmental parameters that co-vary in time, and distance from the source of CO_2. These difficulties combine with natural biological variability within populations to make some types of experimentation impractical. It is however possible to design experiments that minimise the effects of these problems.

Carbon dioxide springs are a resource that will make an important contribution to the scientific debate on the likely effects of global environmental change on ecosystems.

ACKNOWLEDGEMENTS

This research has been supported by the Natural Environment Research Council (GR9-734), Agriculture and Food Research Council (PG16/605), Commission for the European Communities, The British Council, Hotel Terme San Giovanni and LI-COR Inc. Technical assistance was provided by Erica De Leau, John Findlay and Roz Tremlow. The cost of the research has been partially covered by EU, Programme Environment, contract EV5V-CT93-0438.

REFERENCES

Acock, B. (1990) Effects of carbon dioxide on photosynthesis, plant growth, and other processes. In *Impact of carbon dioxide, trace gases, and climate change on global agriculture*, ASA Special Publication, (eds. B.A. Kimball, N.J. Rosenberg, L.H. Allen, G.H. Heichel, C.W. Stuber, D.E. Kissel & S. Ernst), pp. 45-60. American Society of Agronomy, Madison.

Allen, L.H., Drake, B.G., Rogers, H.H, Shinn, J.H. (1992) Field techniques for exposure of plants and ecosystems to elevated CO_2 and other trace gases. *Critical Reviews in Plant Sciences*, 11, 85-119.

Arp, W.J. (1991) Effects of source-sink relations on photosynthetic acclimation to elevated CO_2. *Plant, Cell and Environment*, 14, 869-875.

Bazzaz, F.A. (1990) The response of natural ecosystems to the rising global CO_2 levels. *Annual Review of Ecology and Systematics*, 21, 167-196.

Beerling, D.J. & Chaloner, W.G. (1993a) The impact of atmospheric CO_2 and temperature change on stomatal density observations from *Quercus robur* lammas leaves. *Annals of Botany*, 71, 231-235.

Beerling, D.J. Chaloner, W.G. (1993b) Stomatal density responses of Egyptian *Olea europaea* L. leaves to CO_2 change since 1327 BC. *Annals of Botany*, **71**, 431-435.

Bowes, G. (1993) Facing the inevitable: Plants and increasing atmospheric CO_2. *Annual Review of Plant Physiology and Plant Molecular Biology*, **44**, 309-332.

Crawford, R.M.M. (1989) *Studies in plant survival. Ecological case histories of plant adaptation to adversity*, Blackwell Scientific Publications, Oxford.

Drake, B.G. (1992) The impact of rising CO_2 on ecosystem production. *Water Air and Soil Pollution*, **64**, 25-44.

Hellings, S.E. & Gallagher, J.L. (1992) The effects of salinity and flooding on *Phragmites australis. Journal of Applied Ecology*, **29**, 41-49.

Kimball, B.A. (1992) Cost comparisons among free air CO_2 enrichment, open top chamber, and sunlit controlled environment chamber methods of CO_2 exposure. *Critical Reviews in Plant Sciences*, **11**, 265-270.

Kimball, B.A., Mauney, J.R., Nakayama, F.S. & Idso, S.B. (1993) Effects of increasing atmospheric CO_2 on vegetation. *Vegetatio*, **104**, 65-75.

Lawlor, D.W. & Mitchell, R.A.C. (1991) The effects of increasing CO_2 on crop photosynthesis and productivity - A review of field studies. *Plant, Cell and Environment*, **14**, 807-818.

Leadley, P.W. & Drake, B.G. (1993) Open top chambers for exposing plant canopies to elevated CO_2 concentration and for measuring net gas exchange. *Vegetatio*, **104**, 3-15.

Leuning, R. & Foster, I.J. (1990) Estimation of transpiration by single trees, comparison of a ventilated chamber, leaf energy budgets and a combination equation. *Agricultural and Forest Meteorology*, **51**, 63-86.

Matoh, T., Matsushita, N. & Takahashi, E. (1988) Salt tolerance of the reed plant *Phragmites communis. Physiologia Plantarum*, **72**, 8-14.

McDermitt, D.K., Norman, J.M., Davis, J.T., Ball, T.M., Arkebauer, T.J., Welles, J.M. & Roemer, S.R. (1989) CO_2 response curves can be measured with a field portable closed loop photosynthesis system. *Annales des Sciences Forestières*, **46**, 416-420.

Melillo, J.M., Callaghan, T.V., Woodward, F.I., Salati, E. & Sinha, S.K. (1990) Effects on ecosystems. In *Climatic change. The IPCC scientific assessment*, (eds. J.T. Houghton, G.J. Jenkins & J.J. Ephraums), pp. 283-310. Cambridge University Press, Cambridge.

Miglietta, F., Badiani, M., Bettarini, I., van Gardingen, P.R. & Raschi, A. (1993a). Natural carbon dioxide springs and their use for experimentation. In *Design and execution of experiments on CO_2 enrichment*, (eds. E.D. Schulze & H.A. Mooney), pp. 393-403. Commission of the European Communities, Brussels.

Miglietta, F., Badiani, M., Bettarini, I., van Gardingen, P.R., Selvi, F. & Raschi, A. (1995). Preliminary studies of the long-term CO_2 response of Mediterranean vegetation around natural CO_2 vents. In *Global change and Mediterranean-type ecosystems*, (eds. J.M. Moreno & W.C. Oechel), pp. 102-120. Springer, Berlin.

Miglietta, F. & Raschi, A. (1993) Studying the effect of elevated CO_2 in the open in a naturally enriched environment in Central Italy. *Vegetatio*, **104/105**, 391-400.

Miglietta, F., Raschi, A., Bettarini, I., Resti, R. & Selvi, F. (1993b) Natural CO_2 springs in Italy: A resource for examining long-term response of vegetation to rising atmospheric CO_2 concentrations. *Plant, Cell and Environment*, **16**, 873-878.

Miglietta, F., Raschi, A., Bettarini, I. & Selvi, F. (1992) Photosynthesis and transpiration of *Schoenoplectus tabernaemontani* acclimated to elevated CO_2. Proceedings of the IV International Wetlands Conference, INTECOL, (Abstract).

Parlange, J. & Waggoner, P.E. (1970) Stomatal dimensions and resistance to diffusion. *Plant Physiology*, **46**, 337-342.

Patterson, D.T. & Flint, E.P. (1990) Implications of increasing carbon dioxide and climate change for plant communities and competition in natural and managed ecosystems. In *Impact of carbon dioxide, trace gases, and climate change on global agriculture*, ASA Special Publication, (eds. B.A. Kimball, N.J. Rosenberg, L.H. Allen, G.H. Heichel, C.W. Stuber, D.E. Kissel & S. Ernst), pp. 83-110. American Society of Agronomy, Madison.

Peñuelas, J. & Matamala, R. (1990) Changes in N and S leaf content, stomatal density and specific leaf area of 14 plant species during the last 3 centuries of CO_2 increase. *Journal of Experimental Botany*, **41**, 1119-1124.

Radoglou, K.M. & Jarvis, P.G. (1990) Effects of CO_2 enrichment on 4 Poplar clones. 2. Leaf surface properties. *Annals of Botany*, **65**, 627-632.

Rowland-Bamford, A.J., Nordenbrock, C., Baker, J.T., Bowes, G. & Allen, L.H. (1990) Changes in stomatal density in rice grown under various CO_2 regimes with natural solar irradiance. *Environmental and Experimental Botany*, **30**, 175-180.

van Gardingen, P.R., Grace, J., Harkness, D.D., Miglietta, F. & Raschi A. (1995) Carbon dioxide emissions at an Italian mineral spring: measurements of average CO_2 concentration and air temperature. *Agricultural and Forest Meteorology*, **73**, 17-27.

von Caemmerer, S. & Farquhar, G.D. (1981) Some relationships between the biochemistry of photosynthesis and the gas exchange of leaves. *Planta*, **153**, 376-387.

Weisner, S.E.B., Graneli, W. & Ekstam, B. (1993) Influence of submergence on growth of seedlings of *Scirpus lacustris* and *Phragmites australis*. Freshwater Biology, **29**, 371-375.

Woodward, F.I. (1987) Stomatal numbers are sensitive to increases in CO_2 concentration from pre-industrial levels. *Nature*, **327**, 617-618.

Woodward, F.I. (1992) Predicting plant responses to global environmental change. *New Phytologist*, **122**, 239-251.

Woodward, F.I. & Bazzaz, F.A. (1988) The responses of stomatal density to CO_2 partial pressure. *Journal of Experimental Botany*, **39**, 1771-1781.

Wynn, W.K. (1976) Appressorium formation over stomates by the bean rust fungus: response to a surface contact stimulus. *Phytopathlogy*, **66**, 136-146.

Using Icelandic CO_2 springs to understand the long-term effects of elevated atmospheric CO_2

A. C. COOK[1], W.C. OECHEL[1] AND B. SVEINBJORNSSON[2]

[1] Department of Biology and Global Change Research Group,
San Diego State University, San Diego, CA 92182, USA.
[2] Institute of Biology, University of Iceland, 108 Reykjavik, Iceland
and the University of Alaska, Anchorage, AK 99508, USA.

SUMMARY

The CO_2-emitting mineral springs of Iceland have been surveyed and assessed for their use in understanding the potential effects of elevated atmospheric CO_2 on northern ecosystems. One spring near Olafsvik is described in detail here. This CO_2 spring emits CO_2 in such a way that the surrounding vegetation is exposed to mean CO_2 concentrations which are consistent with those predicted for the next century. Results from carbon isotope analyses show that *Nardus stricta* plants growing near this CO_2 spring are exposed to mean CO_2 concentrations of approximately 880, 650 and 430 ppm, depending on their location relative to the CO_2 vent. Maps and tables are provided to show how the concentration of CO_2 measured in the vegetation varies over time on calm and windy days. A list of the species growing around the Olafsvik CO_2 spring is included and the research potential of CO_2 springs is discussed at length. The use of natural CO_2 springs as surrogates for elevated CO_2 experiments appears highly promising.

INTRODUCTION

The concentration of carbon dioxide in the atmosphere is increasing due to the burning of fossil fuels and deforestation. How plants evolve in response to elevated CO_2 and global climate change will ultimately affect many of the important responses of the biosphere including primary productivity, trace gas flux, and vegetation boundaries. Primarily because of the time-scale of their existence, CO_2 springs (degassing of geologic CO_2 at surface vents or mineral springs) offer a unique opportunity to study the effects of long-term global change on changes in genetic structure, population composition, and ecosystem function, which cannot be conclusively answered using artificial growth conditions in a reasonable period of time. Our goal is to address the immediate need in general circulation and global carbon cycle models for data about the long-term response of terrestrial ecosystems to elevated CO_2. This is accomplished by encouraging the use of natural CO_2 spring studies as surrogates for elevated CO_2 experiments that would otherwise be impractical, expensive, and without results for at least decades to centuries.

How carbon is stored and cycled by plants in northern ecosystems (arctic, boreal forest, and northern bogs) in response to elevated CO_2 and global climate change is of great importance because these ecosystems have the potential to significantly impact the global carbon balance. These ecosystems alone contain about 450 Gt. of carbon (1 gigaton = 10^{15} g), an amount equal to 2/3 of the carbon currently in the atmosphere (Gorham, 1991).

A considerable amount of elevated CO_2 research has already been conducted using arctic species (Tissue & Oechel, 1987; Oberbauer et al., 1986; Billings et al., 1982; 1983; 1984), but we are still far from being able to accurately predict how northern ecosystems will respond to climate change and long-term increases in CO_2. The multivariate nature of natural ecosystems and the multitude of possible interactions and feedbacks contribute to the difficulty of predicting long-term responses. CO_2 springs in Iceland have the potential to provide the long-term data that will be necessary to make more accurate predictions about how northern ecosystems will respond to long-term elevated CO_2 exposure.

We surveyed over 20 of the CO_2-emitting mineral springs in Iceland and were able to identify several with characteristics which would make them particularly suitable for global change research. We were primarily interested in CO_2 springs which (1) were surrounded by native vegetation and (2) emitted CO_2 in such a way that the surrounding vegetation was exposed to levels of CO_2 which are consistent with those predicted for the middle of the next century.

Predictions for the year 2050 are (1) 575 ppm, if emissions grow at a constant relative growth rate of 2% per year, (2) 450 ppm, if emissions remain constant after 1990, and (3) 415 ppm if emissions were reduced by 2% per year from 1990 (Houghton, Jenkins & Ephraums, 1990).

The main objectives of this paper are to (1) briefly describe the CO_2 springs in Iceland, (2) give a detailed description of the CO_2 spring near Olafsvik where we are conducting intensive global change research, and (3) to demonstrate the potential use of natural CO_2 spring "experiments" as surrogates for elevated CO_2 experiments.

ICELANDIC CO_2 SPRINGS

The CO_2 springs of Iceland are concentrated on the Snæfellsnes Peninsula of western Iceland. They can be divided into three groups on the basis of temperature: (1) thermal waters, (2) cold waters with constant temperature, and (3) cold waters which show seasonal variations in temperature (Arnorsson & Barnes, 1983).

The CO_2 springs on Snæfellsnes emit a total of approximately 10 liters per minute of almost pure CO_2 (Å 99%). Since the gas samples collected from the Olafsvik CO_2 spring were contaminated with bulk atmospheric air during transport, we present in Table 1 the composition of the gas being emitted from the Raudimelur CO_2 spring which is representative of the gas being emitted from all of the cold CO_2 springs on Snæfellsnes.

The cold, CO_2-emitting mineral springs of Iceland should not be confused with the hot springs for which Iceland is known. The CO_2 springs of interest are cold, small, surrounded by native vegetation, and of little commercial value. The gas being emitted is free from the sulfur-containing gases which are known to be detrimental to plant growth and are often associated with hot springs.

The gas evolving from the cold water CO_2 spring in Olafsvik was collected and analyzed for hydrogen sulfide, carbon disulfide, dimethyl

TABLE 1. Composition of the gas emitted from the Raudimelur mineral spring on the Snæfellsnes Peninsula of western Iceland. From Arnorsson & Barnes (1983).

Component	Volume %
CO_2	98.72
He	<0.01
H_2	<0.005
Ar	<0.02
O_2	<0.02
N_2	1.10
CH_4	0.02
C_2H_6	<0.01

sulfide, and dimethyl disulfide but all were below the detector's limits of 0.5 ppm for hydrogen sulfide and carbon disulfide and 0.1 ppm for dimethyl sulfide and dimethyl disulfide. Since H_2S can be smelled at 0.025 ppm (National Research Council, 1978) and there is no smell of sulfur present in the gas coming from the CO_2 spring near Olafsvik, there is further evidence for the absence of significant quantities of these phytotoxic gases. Even the most sensitive plants are not affected by hydrogen sulfide until concentrations reach 0.03 ppm (National Research Council, 1978).

Stable carbon isotope composition analyses suggest that the CO_2 emitted from the cold springs in Iceland evolves from the mantle (rather than from decaying organic matter or metamorphism of carbonate minerals) and is transported upward in the gas phase along cracks or fractures in the rock (Arnorsson & Barnes, 1983).

Typically, CO_2 evolving from the mantle has a $\delta^{13}C$ signature of between -4.7 and 8.0 relative to the PDB standard, and according to Arnorsson and Barnes (1983), the CO_2 evolving in Snæfellsnes has a $\delta^{13}C$ value of -5.23. CO_2 which evolves from organic material typically has an isotopic signature of approximately -20 while that evolving from the metamorphism of marine carbonate rocks typically shows a $\delta^{13}C$ value of 0.

We were able to survey most of the known cold CO_2 springs in Iceland during March of 1992 and the summer of 1993. Many of the cold water springs in southern and western Iceland have natural vegetation growing around them and are relatively undisturbed. The only CO_2 springs we did not survey were three located in the polar desert of central Iceland.

These springs were difficult to reach and were unlikely to have had sufficient vegetation surrounding them to support an intensive research program. The CO_2 spring we selected for intensive research is located in a semi-natural grassland near the town of Olafsvik on the Snæfellsnes Peninsula of western Iceland. This CO_2 spring is labeled as number 3 on the map showing the distribution the CO_2 springs in Iceland (Fig. 1).

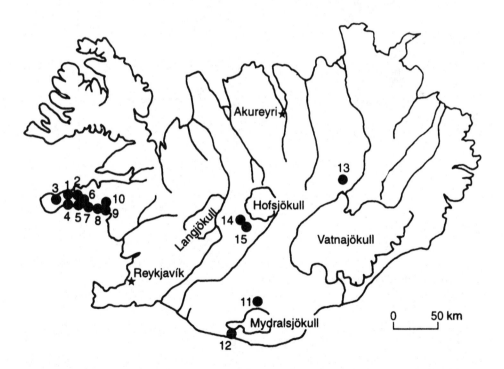

FIG. 1. CO_2-emitting mineral springs in Iceland. 1.Vallnabnikur, 2. Grundarfjordur, 3. Olafsvik, 4. Bjarnarfosskot, 5. Osakot, 6. Lysuholl (bore hole), 7. Lysuholl (swimming pool), 8. Glaumber, 9. Olkelda, 10. Raudimelur, 11. Storibver, 12. Raundafell, 13. Viti, 14. Nedri Hveradalir A and 15. Nedri Hveradalir B. From Holl, Ing & Munzer (1975).

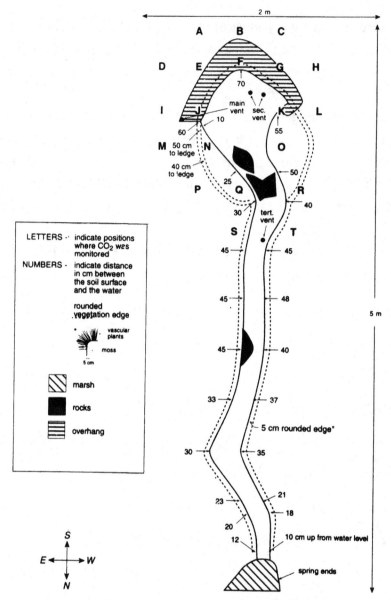

FIG. 2. Map of the CO$_2$-emitting mineral spring near Olafsvik, Iceland

SITE DESCRIPTION

The CO$_2$ spring near Olafsvik has been mapped (Fig. 2). At its maximum, the spring is approximately 1 meter wide and 5 meters long. It is located on a northeast-facing slope, approximately 20 meters above sea level. The level of water in the spring fluctuates with the amount of rainfall and is thought to consist mainly of ground water. The vegetation surrounding the spring does not grow in the spring itself, but rather covers the area above the spring. The vegetation is terrestrial, not aquatic, and there is little change in slope, soil moisture content, or other physical features examined across the CO_2 spring site.

The main CO_2 vent is located at the head of the spring and smaller CO_2 vents can be found along the stream bed. Since CO_2 is denser than air, it has a tendency to be retained within the area of the spring, exposing the surrounding plants to higher levels of CO_2. Reports indicate that this CO_2 vent has been active for at least 110 years and may have been active for much longer (S. Brandsson, personal communication). Isotopic analyses of soil carbon at increasing depths are underway to estimate the date at which CO_2 was first emitted from the spring.

Since a nearby volcano erupted some 2-3 thousand years ago and interrupted the soil record by spreading a thick layer of volcanic ash, the date of first detectable CO_2 emission is not expected to be more than a few thousand years ago (S. Brandsson, personal communication).

One of the most abundant plant species growing in Iceland and around the CO_2 spring in Olafsvik is the perennial pasture grass, *Nardus stricta* (Kristinsson, 1987). Individuals of this species are thought to live for 30-40 years by replacing dying tillers with new tillers. The species is apomictic and reproduces only asexually by vegetative propagation and seed formation without sexual fusion (Perkins, 1968; Chadwick, 1960). This feature may result in a spectrum of specialized, reproductively isolated, co-existing genotypes, including one or more specially adapted to the CO_2-rich environment of the spring. We have chosen to concentrate our efforts on *Nardus stricta*, but a complete list of the species present in a 3 by 5 meter plot around the CO_2 spring has been provided in Table 2. No plants were found growing near the CO_2 spring that could not be found growing elsewhere. While the history of land use around the CO_2 spring near Olafsvik is not precisely known, the town of Olafsvik has been an established place of commerce for the past 300 years. It is likely that the CO_2 spring research area has experienced at least some human

TABLE 2. Inventory of vascular plant species growing around the CO_2 spring near Olafsvik, Iceland. Nomenclature follows that of Kristinsson (1987).

Agrostis capillaris	*Phleum alpinum*
Alchemilla vulgaris	*Pinguicula vulgaris*
Anthoxanthum odoratum	*Plantago maritima*
Bistorta vivipara	*Plantanthera hyperborea*
Caltha palustris	*Poa alpina*
Carex bigelowii	*Poa annua*
Carex nigra	*Potentilla palustris*
Carex serotina	*Ranunculus acris*
Cerastium fontanum	*Rhinanthus minor*
Deschampsia caespitosa	*Rumex acetosa*
Equisetum arvense	*Salix herbacea*
Equisetum fluviatile	*Taraxacum (officinale)*
Eriophorum angustifolium	*Thalictrum alpinum*
Festuca richardsonii	*Tofieldia pusilla*
Festuca vivipara	*Viola palustris*
Fragaria vesca	
Galium verum	The most common mosses are:
Juncus filiformis	*Sphagnum spp.*
Juncus trifidus	*Polytrichum alpinum*
Luzula arcuata	*Polytrichum commune*
Luzula arcuata	*Rhytidiadelphus squarrosus*
Luzula multiflora	*Aylocomium splendens*
Nardus stricta	*Calliergon sp.*

impact and grazing during the past several centuries, but there is no evidence that farming or other very intense activity has occurred in the research area. For at least the past 30 years, the research area has gone essentially unused by the local people. Since livestock have been seen on occasion to enter within 50 to 100 meters of the CO_2 spring research area through imperfections in the exterior fence, we decided to insure that the study area would be protected for future research by enclosing each of the three main study sites (1 CO_2 spring and 2 "control" sites) with electrical fencing and installing a second, more secure, chain-link fence around the perimeter of the area we were specifically interested in. With the additional fences in place, it is now impossible for livestock or people to unintentionally disturb the areas of study. To minimize our own trampling while making measurements, wooden board-walks were installed at each of the 3 main study sites. All of the research plots used in this study (1 CO_2 spring and 5 "control" plots) are located within 50 meters of each other. A schematic for the layout of the research area is provided in Fig. 3.

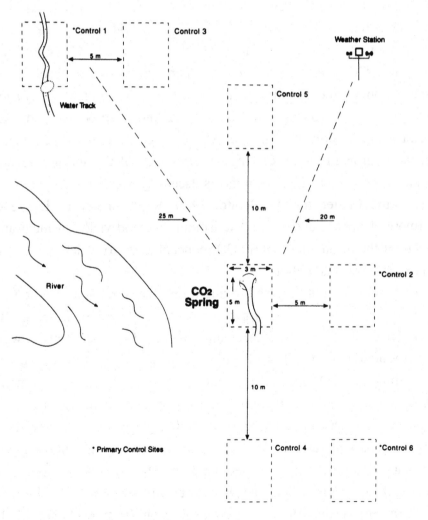

FIG. 3. Schematic for the CO_2 spring research site near Olafsvik, Iceland.

CO_2 MEASUREMENTS

Two methods have been used to characterize the level of CO_2 exposure in the vegetation surrounding the CO_2 spring near Olafsvik.

The first involves collecting plant tissue, having it analyzed for carbon isotope composition (^{12}C, ^{13}C, ^{14}C), and back-calculating to the mean long-term average atmospheric CO_2 concentration by using equations developed by Geyh & Scheider (1990). This method works particularly well at the spring in Olafsvik because the CO_2 emitted from this vent has a distinctive carbon isotope signature.

In simplest form, the CO_2 concentration to which a plant has been exposed can be calculated using: $C_s = C_a * M_a / M_p$ where C_s = mean CO_2 concentration to which a plant has been exposed, C_a = ambient atmospheric CO_2 concentration in parts per million, M_a = absolute ^{14}C concentration of the air minus discrimination (^{14}C content of Nardus stricta plant tissue growing in a "control" site known to be exposed to an ambient concentration of CO_2) and $M_p = ^{14}C$ content of Nardus stricta plant tissue growing in the CO_2-rich environment of the spring (Geyh & Scheider, 1990). The ^{14}C measurements necessary for these analyses were made in the Center for Accelerator Mass Spectrometry at Lawrence Livermore National Laboratory. The isotopic method yields an integrated measure of the mean atmospheric CO_2 concentration to which a plant has been exposed over its lifetime.

The second method involved taking direct measurements of the CO_2 concentration with an infrared gas analyzer (LiCor 6200, Lincoln NE, USA). These short-term measurements of the CO_2 concentration were made sequentially at 20 grid points positioned in the vegetation surrounding the CO_2 spring. Typically, 7-8 sets of 10 measurements each were calculated per grid point per day over a 16 hour period (for a total of 70 or 80 measurements recorded per grid point per day). Each measurement recorded by the computer of the gas analyzer was programmed to represent a ten second average CO_2 concentration (the average of 10 single second instantaneous measurements of the CO_2 concentration). Using the mean concentration from each set of 10 second measurements, average CO_2 concentrations were calculated for each grid point and plotted to yield a map of the plume of CO_2 present in the vegetation surrounding the CO_2 spring. These measurements were taken on one calm and one windy day.

RESULTS

Using carbon isotope measurements and the equations from Geyh & Scheider (1990), we found that the plants in the areas sampled were exposed to mean CO_2 concentrations of 874 ± 27, 650 ± 20, and 432 ± 11 ppm (mean $\pm$ SE, where n = 3, 3, and 2 samples respectively and each sample represents a random 300 mg portion of the plant material that was collected at different distances from the CO_2 spring and pooled). The approximate locations for these collection sites are shown in Fig. 2 (as J, K, and F for 874 ppm, as N, Q, and R for 650 ppm and as the rock labeled 45 for 432 ppm).

On August 21 and 22 of 1992, the plume of CO_2 in the vegetation surrounding the Olafsvik CO_2 spring was measured in detail using an infrared gas analyzer. The average CO_2 concentration calculated for each grid point is depicted in Fig. 4 using a gray scale where darker shades indicate higher concentrations of CO_2 and lighter shades indicate lower concentrations of CO_2.

Fig. 4a represents the distribution of CO_2 in the vegetation on a calm day (slight breeze out of the NW at 1.1 m/s), while Fig. 4b represents the plume of CO_2 in the vegetation on a windy day (wind out of the NW at an average speed of 3.7 m/s).

The concentration of CO_2 measured in the vegetation ranged from the just above ambient (355 ppm) to more than 1000 ppm. A table has been provided to show the range and variability of the CO_2 concentrations measured at the CO_2 spring on both calm and windy days (Table 3).

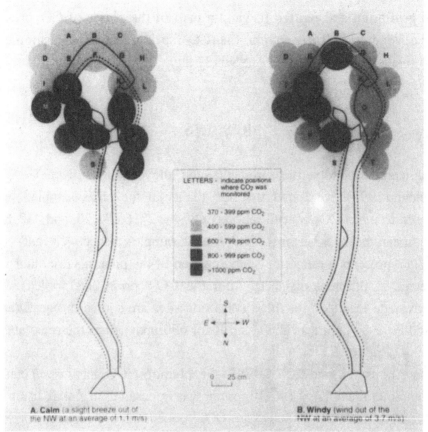

FIG. 4. Distribution of CO$_2$ on a calm and a windy day at Olafsvik, Iceland.

DISCUSSION

The CO$_2$ concentration in the vegetation surrounding the CO$_2$ spring near Olafsvik has been shown to vary greatly over time depending on wind speed and direction. While it is normal for plants to experience at least some seasonal and diurnal variation in CO$_2$ concentration, the extent of variation near CO$_2$ springs is large and may be greater than that expected under future global change scenarios. Isotopic measurements provided a reliable method of measuring the long-term integrated lifetime mean CO$_2$ concentration to which the plants in the field have been exposed, while gas analyzer measurements provided some estimation of the short-term variability in atmospheric CO$_2$ concentrations. The effects of short-term variation in CO_2 concentration on plant growth and physiology are not yet known, but should be kept in mind when interpreting the results from around CO_2 springs. The lifetime mean CO_2 exposure concentrations that were calculated using the isotopic method were consistent with those predicted for the middle of the next century, but it is unclear whether the mean CO_2 concentration or its variability is more important.

TABLE 3. CO_2 concentration and variability in the vicinity of the Olafsvik CO_2 spring.

Grid point	Mean [CO_2] ± Standard deviation				Range in ppm CO_2	
	calm	n	windy	n	calm	windy
A	419±24	7	373±12	7	388-454	364-396
B	405±18	7	390±20	7	383-433	361-423
C	426±38	8	398±33	7	368-489	370-466
D	429±23	7	439±102	8	393-460	365-646
E	532±50	7	487±75	7	462-602	392-625
F	570±76	7	612±151	7	481-693	478-852
G	457±71	7	430±22	7	388-568	408-469
H	432±38	7	397±13	7	370-477	380-420
I	515±153	7	484±65	6	410-855	383-555
J	899±316	7	1201±237	7	640-1540	855-1528
K	599±67	7	492±63	8	522-733	412-606
L	480±33	7	403±20	7	445-523	382-442
M	932±513	7	749±205	7	539-1850	516-1026
N	1785±544	7	1293±308	6	883-2319	949-1837
O	1020±385	7	472±68	7	634-1662	398-567
P	719±213	7	418±35	7	483-1142	362-454
Q	1450±522	22	814±270	21	588-2314	441-1250
R	981±581	7	421±39	7	423-1928	385-481
S	518±118	7	387±24	7	418-752	358-426
T	958±326	6	404±32	7	586-1564	376-454
AMBIENT	363±21	2			348-378	

The main advantage of using CO_2 rich mineral springs in ecophysiological studies is that the atmosphere surrounding them has been naturally enriched for hundreds of years or more, giving the plants in the vicinity of these springs enough time to evolve under conditions of elevated CO_2. The extended duration of elevated CO_2 exposure available near CO_2 springs makes it possible to study the long-term plant and ecosystem responses that can take several centuries to manifest. Since CO_2 springs expose intact ecosystems to elevated CO_2, community level processes requiring competition over many generations can finally be investigated. Research around CO_2 springs addresses the immediate need in general circulation and global carbon cycle models for data about the long-term response of terrestrial ecosystems to elevated CO_2.

While modeling efforts based on shorter-term CO_2 enrichment studies may provide some insight into the long-term response of vegetation and ecosystems to rising CO_2, it remains unclear whether extending the length of elevated CO_2 exposure to centuries will change the trajectory of the response found in short-term experiments. Some have argued that compensatory mechanisms acting over longer time-scales (changes in nutrient cycling, sink strength, enzyme efficiency, phenology, adaptation, etc.) will alter the trajectory of the response found in shorter-term CO_2 enrichment studies. Studies around CO_2 springs are expected to clarify some of this uncertainty by providing the long-term data necessary to test whether models based on short-term data can reliably predict long-term results. CO_2 springs offer one means of such model validation.

Research around CO_2 springs in Iceland will contribute directly to our understanding of how northern ecosystems will respond to long-term elevated CO_2 exposure. These ecosystems alone contain 25-33% of the global soil carbon pool (Billings, 1987) and have the potential to significantly impact the global carbon cycle and feedback on CO_2-induced climate change. Recent studies suggest that the carbon balance of northern ecosystems is changing (Grulke *et al.*, 1990; Oechel *et al.*, 1993), but how these changes will manifest themselves over longer time scales is not known. CO_2 springs in Iceland should provide significant insight into the long-term response of northern ecosystems to elevated CO_2.

Although each CO_2 spring study will provide valuable information about how a specific ecosystem will respond to long-term elevated CO_2 exposure, surveying CO_2 springs located in various ecosystems will allow us to look for patterns in the response of different ecosystems to elevated CO_2. If trends emerge between different ecosystems, it could lead to drastic improvements in our ability to predict how terrestrial ecosystems will respond to, and feedback upon, future concentrations of atmospheric CO_2. Every effort should be made to find at least one replicate CO_2 spring for each ecosystem type, so as to permit more reliable comparison of the response of different ecosystems to long-term elevated CO_2 exposure.

Even though CO_2 spring studies are purely observational and not controlled manipulations, we propose the use of them as surrogates for long-term elevated CO_2 experiments because they have the potential to provide information immediately that would otherwise require intense effort and centuries to obtain. CO_2 springs provide a unique opportunity to investigate long-term global change questions that are simply outside the scope and duration of most experimental research and are likely to remain unanswered if only traditional methods are used.

ACKNOWLEDGEMENTS

We are grateful to the people of Olafsvik, Iceland for their gracious support of this research effort and to the Center for AMS at Lawrence Livermore National Laboratory for making the ^{14}C isotope measurements. We also thank several reviewers for critically reading the earlier versions of this manuscript. This research was supported by the Department of Energy grant DE-FG03-86ER60479, a doctoral enhancement award from the National Science Foundation INT-9224941, the University of Iceland Research Foundation, the Joint Doctoral Program in Ecology at San Diego State University and the University of California at Davis, Sigma Xi, and awards from the San Diego Chapter of Achievement Rewards for College Scientists, Inc.

REFERENCES

Arnorsson, S. & Barnes, I. (1983) The nature of the carbon dioxide waters in Snæfellsnes, Western Iceland. *Geothermics*, **12**, 171-176.

Billings, W.D (1987) Carbon balance of Alaskan tundra and taiga ecosystems: past, present, and future. *Quart. Sci. Rev.*, **6**, 165-177.

Billings, W.D., Luken, J.O., Mortensen, D.A. & Peterson, K.M. (1982) Arctic tundra: A sink or source for atmospheric carbon dioxide in a changing environment? *Oecologia,* **53,** 7-11.

Billings, W.D., Luken, J.O., Mortensen, D.A. & Peterson, K.M. (1983) Increasing atmospheric carbon dioxide: possible effects on arctic tundra. *Oecologia,* **58,** 286-289.

Billings, W.D., Luken, J.O., Mortensen, D.A. & Peterson, K.M. (1984) Interaction of increasing atmospheric carbon dioxide and soil nitrogen on the carbon balance of tundra microcosms. *Oecologia,* **65,** 26-29.

Chadwick, M.J. (1960) Biological Flora of the British Isles: *Nardus stricta. Journal of Ecology,* **48,** 255-267.

Geyh, M.A. & Scheider, H. (1990) *Absolute Age Determination.* Springer-Verlag, Berlin.

Grulke, N.E., Riechers, G.H., Oechel, W.C., Hjelm, U. & Jaeger, C. (1990) Carbon balance in tussock tundra under ambient and elevated atmospheric CO$_2$. *Oecologia,* **83,** 485-494.

Gorham, E. (1991) Northern peatlands: Role in the carbon cycle and probable responses to climatic warming. *Ecological Applications,* **1,** 182-195.

Holl, K., Ing, E.H. & Munzer, U. (1975) Islands Olkelda (Mineralquellen). *Hveragerdi,* **19,** 1-50 (in German).

Houghton, J.T., Jenkins, G.J., & Ephraums, J.J. (1990) (eds.) *Climate Change: The IPCC Scientific Assessment.* Cambridge University Press, Cambridge.

Kristinsson, H. (1987) *A Guide to the Flowering Plants and Ferns of Iceland.* Orn and Orlygur Publishing House, Reykjavik.

National Research Council, Committee on Medical and Biologic Effects of Environmental Pollutants, Subcommittee on Hydrogen Sulfide (1978) *Hydrogen Sulfide,* University Park Press, Baltimore.

Oberbauer, S.F., Sionit, N., Hastings, S.J. & Oechel, W.C. (1986) Effects of CO$_2$ enrichment and nutrition on growth, photosynthesis, and nutrient concentration of Alaskan tundra plant species. *Canadian Journal of Botany,* **64,** 2993-2998.

Oechel, W.C., Hastings, S.J., Vourlitis, G.L., Jenkins, M., Riechers, G. & Grulke, N. (1993) Recent change of arctic tundra ecosystems from a net carbon dioxide sink to a source. *Nature* **361,** 520-523.

Perkins, D.F. (1968) Ecology of *Nardus stricta* L. Annual growth in relation to tiller phenology. *Journal of Ecology,* **56,** 633-646.

Tissue, D.L. & Oechel, W.C. (1987) Physiological response of *Eriophorum vaginatum* to elevated CO$_2$ and temperature in the Alaskan tussock tundra. *Ecology,* **68,** 401 - 410.

Plant-CO_2 responses in the long term: plants from CO_2 springs in Florida and tombs in Egypt

F.I. WOODWARD AND D.J. BEERLING

Department of Animal and Plant Sciences,
University of Sheffield, S10 2TN, UK.

SUMMARY

Seeds were collected from populations of *Boehmeria cylindrica* growing in naturally enriched concentrations of CO_2. In the controlled environment plants grown from populations which are normally exposed to enriched atmospheres of CO_2 (400 to 575 ppmv) show a greater CO_2-sensitivity of growth than plants from ambient (370 ppmv) sites. Observed differences in height growth and dry weight partitioning could be explained in terms of the different rates of plant growth under the imposed treatments, rather than in terms of population differences in growth partitioning. Comparisons of stomatal density-CO_2 responses for species from different climates and over the different time scales of centuries and millennia indicate a similar decline in density with increasing CO_2 concentrations, with no evidence of any changes in the CO_2 sensitivity. However, the clear species-specific nature of the stomatal density response suggests it is unlikely to be observed ubiquitously in plants growing by high-CO_2 springs. Nevertheless, there is a close similarity between the stomatal density responses shown by plants growing under CO_2 enrichment adjacent to springs, and the stomatal density changes recorded from fossil leaves, indicating one valuable use of high-CO_2 springs in biological research.

INTRODUCTION

Measurements of the atmospheric concentration of CO_2, typically from air locked up in permanent ice caps, provide time courses for a continuously varying CO_2 concentration over the last 160,000 years (Barnola *et al.*, 1987). Estimates of CO_2 by indirect means also show variations of CO_2 concentration over millions of years (Hays, Imbrie & Shackleton, 1976; Berner, 1990).

An interesting consequence of these variations is the potential for a CO_2-controlled selection in higher plants. As the time scales of CO_2 variations would accommodate many life cycles, even for long-lived plants, then there is a real possibility for selection, if there is an adequate selective force due to CO_2 and if there is adequate variation, within plant populations, in CO_2 responsiveness. Work on *Plantago lanceolata* (Wulff & Alexander, 1985) and *Phlox drummondii* (Garbutt & Bazzaz, 1984) has demonstrated significant genetic variation in CO_2 responses, indicating a potential for long term selection. This paper investigates the probability of CO_2-related selection by two different approaches.

The first approach analyses the growth responses to CO_2 enrichment by plants collected from different populations, at sites differing naturally in ambient CO_2 concentrations (Woodward, Thompson & McKee, 1991; Woodward, 1993). The hypothesis to be tested is that populations growing under different ambient concentrations of CO_2 will show differential responses to CO_2 enrichment under controlled conditions of growth.

The second approach addresses the question of plant-CO_2 responses by determining the responsiveness of stomatal density to changes in CO_2 over time scales of decades, centuries and millennia. This work emerges from the observations by Woodward (1987) that stomatal density decreases in response to an increase in atmospheric CO_2 concentration. Subsequent work by Beerling (Beerling & Chaloner, 1992; Beerling, 1993; Beerling & Chaloner, 1993a, 1993b, 1993c; Beerling *et al.*, 1993) has demonstrated that macrofossil leaves, at least up to 140,000 years old, can also provide quantitative information on stomatal density.

The hypothesis to be tested here is that the response of stomatal density to variations in CO_2 concentrations may be independent of the climate experienced by plants from different regions of the globe and that the response to CO_2 has not changed with time.

This understanding is critical when using the CO_2 response of stomatal density as an indicator of historical concentrations of CO_2 over long time sequences (e.g. Van der Burgh *et al.*, 1993).

MATERIALS AND METHODS

Plants from sites of natural CO_2 enrichment

Seeds from populations of the annual species *Boehmeria cylindrica* were collected from the edges of cold springs in Florida, USA (Woodward *et al.*, 1991; Woodward, 1993).

These sites can be enriched in CO_2 as a result of out-gassing from the cold springs (Rosenau *et al.*, 1977).

Seeds from two different populations growing in enriched CO_2 concentrations (450 to 575 ppmv) and from two different populations growing under ambient CO_2 concentrations (370 ppmv) were collected and subsequently grown in controlled environments, at different CO_2 concentrations (Woodward *et al.*, 1991; Woodward, 1993).

Plants were grown from seedlings and harvested after 54 days, to provide measurements of plant growth (height, dry weight, leaf area) and development (stomatal density and leaf nitrogen concentration).

Observations of stomatal density on herbarium and tomb leaves

Collections of leaves stored in herbaria provide an historical record of stomatal density over the same time span (centuries) as the Siple Station ice core.

This high resolution record of atmospheric CO_2 concentrations (Friedli *et al.*, 1986) can be used to make correlative tests with stomatal distribution.

Here, we have assembled data from prior stomatal observations of the following three species from different climates: *Salix herbacea* (arctic-alpine) (Beerling *et al.*, 1993), *Quercus robur* (temperate tree species)

(Beerling & Chaloner, 1993b) and *Olea europaea* (Mediterranean) (Beerling & Chaloner, 1993c).

Material for the first two species spans the last two centuries, whilst that of *O. europaea* extends this time scale by an order of magnitude using archaeobotanical remains found within King Tutankhamun's tomb, dating from 1327 BC.

Atmospheric CO_2 concentration for this time interval were taken from the Vostok ice core (Barnola *et al.*, 1987).

RESULTS

Analysis of growth and development by Boehmeria cylindrica

Three populations of *Boehmeria* were grown from the seedling stage, one from a site with no CO_2 enrichment (c. 350 ppm), described as the current CO_2 site and two from sites with CO_2 enrichment (c. 450 to 550 ppm), described as the site with elevated CO_2.

The growth responses of the two populations from the enriched sites were not significantly different and have been combined together in the subsequent analysis. The plants from the two populations were arranged as matched pairs, and the mean value of each pair is used in further analyses.

Total plant dry weight after 54 days increased with the CO_2 concentration during growth, with a greater response by the populations from sites with natural CO_2 enrichment (Fig. 1).

The slope of the growth response to CO_2 was significantly (P=0.1) greater for the populations from the elevated CO_2 sites. Total plant height (Fig. 2) also increased, although the greater response by the populations from the elevated CO_2 sites was not significant.

The ratio of shoot to root dry weight, at the end of the experimental period, showed a marked divergence of CO_2 response (Fig. 3) with significant population differences observed at the growth CO_2 mole fraction of 525 µmol mol^{-1}. The shoot to root ratio (Fig. 3) is calculated at the end of the growth experiment. If there is any ontogenetic response of the shoot to root ratio, then it is possible that some of the observed

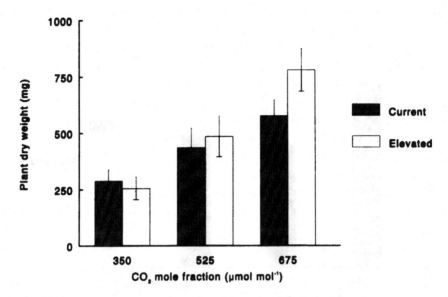

FIG. 1. Total plant dry weight after 54 weeks of growth at three CO_2 mole fractions and for populations collected from sites with current and elevated CO_2 concentration. Means shown with standard errors (n=14). Two populations from sites with CO_2 enrichment (ca. 450 to 550 ppm) are described as the site with elevated CO_2. The growth responses of the two populations from the enriched sites were not significantly different and have been combined together in the subsequent analysis. The plants from the two populations were arranged as matched pairs, and the mean value of each pair is used in further analyses.

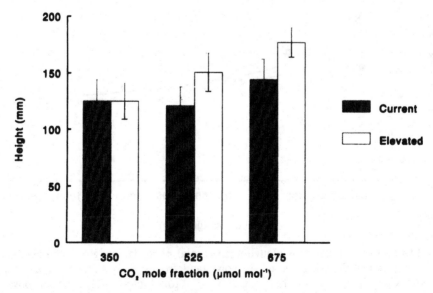

FIG. 2. Response of total plant height to CO_2 mole fraction for populations of *Boehmeria* collected from sites of current and elevated CO_2 concentrations. Symbols as for Fig. 1.

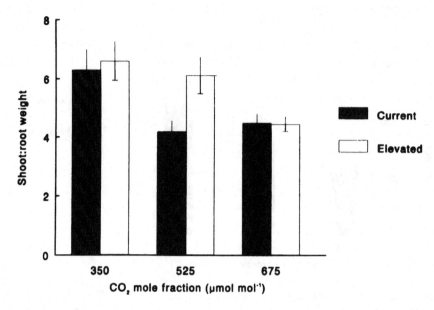

FIG. 3. The response of the shoot to root dry weight ratio to CO_2 mole fraction. Symbols as for Fig. 1.

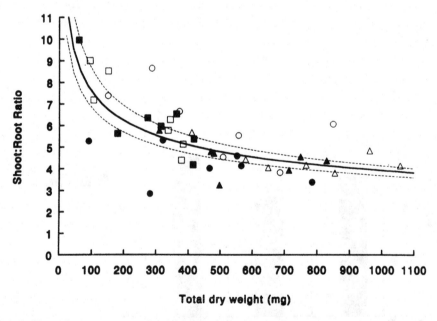

Fig. 4. Plant shoot to root ratios plotted against total plant dry weight. Open symbols plants from elevated CO_2 sites, closed symbols plants from current CO_2 sites, (n, 350 μ mol mol^{-1} CO_2; 1, 525 μmol mol^{-1} CO2; s, 675 μmol mol^{-1} CO_2.) Upper power regression line for elevated CO_2 populations, lower power regression line for current CO_2 populations, central power regression line for both populations. No significant differences between the constants and the coefficients for any of the regression lines.

differences in the ratio are merely a reflection of the different stages of development of the plants in the different treatments. This response has been crudely addressed by plotting the shoot to root ratio against the total dry weight of individual, or paired plants. In this instance total dry weight is used as a measure of the stage of plant development (Fig. 4).

There were no significant differences between the populations when the shoot to root ratio was plotted against total plant dry weight. This response indicates that the observed population differences at the end of the experiment were due to the different rates of plant growth at the different CO_2 mole fractions.

The CO_2 responses of stomatal density (Fig. 5) and leaf nitrogen (data not shown) were also investigated. In both cases there were no significant treatment or population effects, although both populations showed a non-significant decline in stomatal density with increasing CO_2 concentration (Fig. 5).

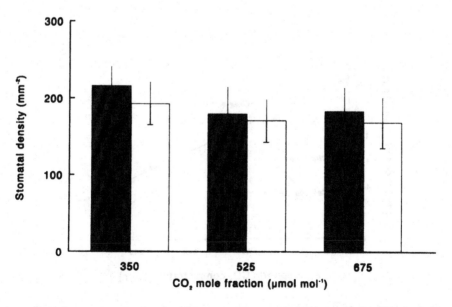

FIG. 5. Stomatal density responses to CO_2 enrichment. Symbols as for Fig. 1.

Analyses of stomatal density responses to historical increases in concentration

Overall the comparison between species originating from different climates shows similar gradients of CO_2 responses (Fig. 6). This is despite differences in the time span between the temperate and arctic-alpine species (centuries) and *Olea europaea* from the Mediterranean (millennia). Statistical comparisons of the regression lines given in Fig. 6 indicate no significant (P<0.05) differences between the slopes of the three lines.

DISCUSSION

The growth analysis of the populations of *Boehmeria* from sites differing in ambient CO_2 concentration demonstrated little evidence for population differences in growth responses. The only clear case was that of the rate of dry weight accumulation (Fig. 1) which was significantly more sensitive to

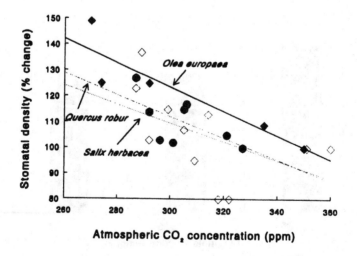

FIG. 6. Comparison of the stomatal density responses of a temperate species (*Quercus robur*, ◊), an arctic-alpine species (*Salix herbacea*, ♦) and a Mediterranean species (*Olea europaea*, •). Slope and r values for each regression line were: -0.36, 0.508; -0.42, 0.613 and -0.46, 0.904 respectively

CO_2 for the population growing in the naturally enriched, elevated CO_2 environment. Measurements of plant gas exchange were not made, however it is most likely that the observed differences in growth were related to population differences in photosynthesis and respiration (Norby *et al.*, 1992; Wullschleger & Norby, 1992). Further studies are clearly necessary to investigate these responses.

Measurements of growth, and growth-partitioning at the end of the experimental period have shown population differences in CO_2 responses (e.g. Fig. 3), however these differences can be most easily explained as a consequence of the different rates of growth and development with CO_2 enrichment (Fig. 4; Coleman, McConnaughay & Bazzaz, 1993). In conclusion, the growth study has demonstrated population differences in CO_2-dependent growth rates which are correlated with the natural CO_2 concentrations in which the plants are grown. Population differences in growth partitioning and height growth were recorded, but these can be most easily explained as a consequence of the different rates of growth and development.

The stomatal density responses to historical increases in the concentration of atmospheric CO_2 appear to be consistent with species of differing climates, indicating the relative robustness of extrapolating to older geological periods (c.f. Van der Burgh *et al.*, 1993). Previous comparisons of larger numbers of species from both climates (Woodward, 1993) suggested that Mediterranean species exhibited a decline somewhat less than that of temperate species, although in both cases the greatest decrease occurred before the 20th century.

Overall, however, where species show a reduction in stomatal density with CO_2 concentration, the degree of the response appears similar between climates and over different periods of time. The latter observations suggest a constancy in the response, even though time has been available for the selection of a changed response. However, species do differ in the stomatal density responses to CO_2, but it has not yet been possible to analyse sufficient data to explain these differences as responses due to selection, or due to carry-over from a particular, perhaps random response by an ancestral species.

There is now some evidence to suggest that stomatal density-CO_2

responses are species-specific with some species showing a decline above present-day CO$_2$ concentrations (e.g. Kelly, Hicklenton & Reekie, 1991) and others showing no response (e.g. Radoglou & Jarvis, 1990).

This species-specific nature of the response may explain the slight reduction in stomatal density of populations of *Boehmeria* growing at increasingly greater CO$_2$ concentrations (Fig. 5). It contrasts with the significant reduction in density seen in *Quercus ilex* and other species growing near elevated CO$_2$ springs, as described elsewhere in this volume.

One interesting feature of high-CO$_2$ springs in Iceland is the occurrence of *Salix herbacea* both at sites of low and naturally high CO$_2$ concentrations. The stomatal density response of *S. herbacea* at these sites, a decline with increasing CO$_2$ concentration (B. Sveinbjornsson, *pers. comm.*), independently confirms that reported from the fossil record during the last glacial-interglacial cycle over which time CO$_2$ concentrations increased from c. 180 ppm to 280 ppm (Barnola *et al.*, 1987). In both cases *S. herbacea* was exposed naturally to changes in the concentration of atmospheric CO$_2$, but in very different situations, yet responded in a consistent manner. This example serves to highlight the usefulness of high-CO$_2$ springs in drawing together independent lines of evidence for plant responses to past and future global change.

REFERENCES

Barnola, J.M., Raynaud, D., Korotkevich, Y.S. & Lorius, C. (1987) Vostok ice core provides 160,000 years record of atmospheric CO$_2$. *Nature*, **329**, 408-414.

Beerling, D.J. (1993) Changes in the stomatal density of *Betula nana* leaves in response to increases in atmospheric carbon dioxide concentration since the late-glacial. *Special Papers in Palaeontology*, **49**, 181-187.

Beerling, D.J. & Chaloner, W.G. (1992) Stomatal density as an indicator of atmospheric CO$_2$ concentration. *The Holocene* 2, 71-78.

Beerling, D.J. & Chaloner, W.G. (1993a) Evolutionary responses of stomatal density to global CO$_2$ change. *Biological Journal of the Linnean Society*, **48**, 343-353.

Beerling, D.J. & Chaloner, W.G. (1993b) The impact of atmospheric CO$_2$ and temperature change on stomatal density: observations from *Quercus robur* lammas leaves. *Annals of Botany*, **71**, 231-235.

Beerling, D.J. & Chaloner, W.G. (1993c) Stomatal density responses of Egyptian *Olea europaea* L. leaves to CO_2 change since 1327 BC. *Annals of Botany,* **71,** 431-435.

Beerling, D.J., Chaloner, W.G., Huntley, B., Pearson, J.A. & Tooley, M.J. (1993) Stomatal density responds to the glacial cycle of environmental change. *Proceedings of the Royal Society of London Series,* **B 251,** 133-138.

Berner, R.A. (1990) Atmospheric carbon dioxide levels over Phanerozoic time. *Science,* **249,** 1382-1392.

Coleman, J.S., McConnaughay, K.D.M. & Bazzaz, F.A. (1993) Elevated CO_2 and plant nitrogen-use: is reduced tissue concentration size dependent? *Oecologia,* **93,** 195-200.

Friedli, H., Lötscher, H., Oeschger, H., Siegenthaler, U. & Stauffer, B. (1986) Ice core record of the $^{13}C/^{12}C$ ratio of atmospheric CO_2 in the past two centuries. *Nature,* **324,** 237-238.

Garbutt, K. & Bazzaz, F.A. (1984) The effects of elevated CO_2 on plants. III. Flower, fruit and seed production and abortion. *New Phytologist,* **97,** 433-446.

Hays, J.D., Imbrie, J. & Shackleton, N.J. (1976) Variation in the Earth's orbit: pacemaker of the Ice Ages. *Science,* **194,** 1121-1132.

Kelly, D.W., Hicklenton, P.R. & Reekie, E.G. (1991) Photosynthetic response of geranium to elevated CO_2 as affected by leaf age and time of CO_2 exposure. *Canadian Journal of Botany,* **69,** 2482-2488.

Norby, R.J., Gunderson, C.A., Wullschleger, S.D., O'Neill, E.G. & McCracken, M.K. (1992) Productivity and compensatory responses of yellow poplar trees in elevated CO_2. *Nature,* **357,** 322-324.

Radoglou, K.M. & Jarvis, P.G. (1990) Effects of CO_2 enrichment on four poplar clones. II. Leaf surface properties. *Annals of Botany,* **65,** 627-632.

Rosenau, J.C., Faulkner, G.L., Hendry, C.W. & Hull, R.W. (1977) *Springs of Florida.* Florida Department of Natural Resources, Tallahassee, USA.

Van der Burgh, J., Visscher, H., Dilcher, D.L. & Kürschner, W.M. (1993) Paleoatmospheric signatures in Neogene fossil leaves. *Science,* **260,** 1788-1790.

Woodward, F.I. (1987) Stomatal numbers are sensitive to increases in CO_2 from pre-industrial levels. *Nature,* **327,** 617-618.

Woodward, F.I. (1993) Plant responses to past concentrations of CO_2. *Vegetatio,* **104/105,** 145-155.

Woodward, F.I., Thompson, G.B. & McKee, I.F. (1991) The effects of elevated concentrations of carbon dioxide on individual plants, populations, communities and ecosystems. *Annals of Botany,* **67** (Supplement 1), 23-38.

Wulff, R.D. & Alexander, H.M. (1985) Intraspecific variation in the response to CO_2 enrichment in seeds and seedlings of *Plantago lanceolata* L. *Oecologia,* **66,** 458-460.

Wullschleger, S.D. & Norby, R.J. (1992) Respiratory cost of leaf growth and maintenance in white oak saplings exposed to atmospheric CO_2 enrichment. *Canadian Journal of Forest Research,* **22,** 1717-1721.

Acidophilic grass communities of CO$_2$-springs in central Italy: composition, structure and ecology

F. SELVI

Dipartimento di Biologia Vegetale, Università di Firenze,
Via la Pira 4, I-50121, Firenze, Italy.

SUMMARY

The results of a study on the composition, the ecology and the structure of grass communities developed close around six mineral CO$_2$-springs in central-western Italy are reported. The phytosociological sampling of the grasslands surrounding the gas vents has led to the circumscription of an azonal endemic association, the *Agrostidetum caninae* subsp. *monteluccii*, which was characterized by a strong species poorness and a high ecological specialization. The typical monospecific stands were developed on peat-like soils with pH ranging between 2.4 and 3.7 and a content of soluble aluminium well above the toxicity thresholds known within the Al-tolerant gen. *Agrostis*. On the basis of published data and preliminary measurements of atmospheric CO$_2$ concentration, it is known that this association is developed in areas with elevated levels of atmospheric CO$_2$, a factor which could contribute to enhancing the competitive capacity of its dominant species within the natural vegetation. Indeed, above-ground biomass, canopy height and other structural parameters of *A. canina* revealed its outstanding vegetative vigour in the monospecific grassland close around the vents. On the contrary, in the peripheral areas of the CO$_2$ springs a decreased vigour in *A. canina* accounted for the different structure and density observed in the grass community and probably also for its different species composition. The syntaxonomical placing of the Agrostidetum is not immediate because of the absence in the European literature of other works on geothermal acidophilic vegetation. Therefore, despite some ecological affinities with oligotrophic species-poor communities of European acidic fens and mires of the Scheuchzerio-Caricetea fuscae, the syntaxonomical placing requires further coenological investigation of geothermal sites on a broader geographical scale.

INTRODUCTION

The western regions of peninsular Italy are very rich in surface geothermal manifestations, connected to residues of volcanic activity or to the presence of deep consolidating magmas with a relatively high radioactivity (T.C.I., 1957). It is not surprising, therefore, that more than 110 gas springs exist in the magmatic districts of central-western Italy (Panichi & Tongiorgi, 1975; Duchi, Minissale & Romani, 1985) as for example around Monte Amiata in southern Tuscany and Monti Cimini in northern Latium (Sollevanti, 1983). In these areas, deep geothermal anomalies frequently cause surface phenomena, like hot water springs and gaseous emissions (Baldi et al. 1974). In most of the cases, the emitted gas is mainly composed of carbon dioxide (90-97 % by volume), small amounts of methane (2.6%), nitrogen (5.2%) and hydrogen sulphide (traces) (Duchi et al., 1985). Although most CO_2 spring sites are currently exploited for industrial production of sparkling waters, so that the natural vegetation is often largely altered by human activities, some still exist in which the spatial arrangement and species composition of plant communities around the vents are undisturbed.

Since the high gas emission rate of the vents is known to produce a significant enhancement of the atmospheric CO_2 concentration in the proximity of some of these sites (Miglietta & Raschi, 1993; Miglietta et al., 1993a), the interest of the scientific community has recently focused on the possibility of using CO_2 springs to investigate the long-term effects of elevated CO_2 on natural vegetation. However, the ecological studies in gas springs (Woodward, Thompson & Mc Kee, 1991; Miglietta & Raschi, 1993; Miglietta et al., 1993a; Miglietta et al., 1993b; Koch, 1994; Körner & Miglietta, 1994) were mostly focused on the physioecological response of the individual plant to elevated CO_2, without paying any attention to the phytocoenological and synecological aspects of the pioneer vegetation colonizing the surroundings of these peculiar habitats. Indeed, a knowledge of these aspects can provide a basis for further investigations, especially at the community level, as well as circumstantial evidence of other "disturbing" factors controlling the species richness, the structure and the patterns of spatial arrangement of the vegetation.

Preliminary observations (Miglietta *et al.*, 1992; Miglietta *et al.*, 1995) showed that different types of plant communities tended to surround the gas vents following the model of the "inclusive niches" proposed by Wisheu & Keddy (1992). The observed consistency in the directional zonation of these "niches" around several widely separated CO_2 springs was supposed to be controlled by a mechanism of competitive hierarchy, in which the weaker competitors were restricted to the peripheral end of an environmental gradient as a result of a trade-off mechanism between tolerance limits and competitive abilities.

This paper is focused on the "grassland niche", and describes some aspects concerning its composition, ecology and structure on the basis of data collected in six geothermal CO_2 springs in southern Tuscany and northern Latium.

MATERIALS AND METHODS

The six geothermal sites with natural CO_2 emissions were selected for the low degree of human disturbance of the surrounding vegetation; they are (Fig. 1): Castiglioni (CA), Vagliagli (VA), Carpineto (CV), Solfatara (SV), Caldara (CM) and Monterano (CB). The first three are located in southern Tuscany (province of Siena) and the others in northern Latium (province of Viterbo and Rome). All are situated in hilly areas with an altitude between 200 and 470 m a.s.l., the Tuscan in the evergreen vegetation area and the Latial in that of deciduous vegetation. The largest of these gas springs was Caldara di Manziana, a crateric depression with a diameter of about 800 m; the others were much smaller, with a diameter ranging between 40 and 150 m. The morphology of the gas springs was flat and rich in hollows and depressions from which gases and cold bubbling waters issued.

Firstly, the herbaceous communities surrounding the gas vents of the six sites were subjectively sampled by means of 31 phytosociological sample plots. The resulting floristic data matrix was then processed by means of the programs Hmcl and Princomp of the Syntax IV package (Podani, 1991), after the transformation of the Braun-Blanquet cover values into the ordinal scale of Westhoff & Van der Maarel (1978). The eight plots

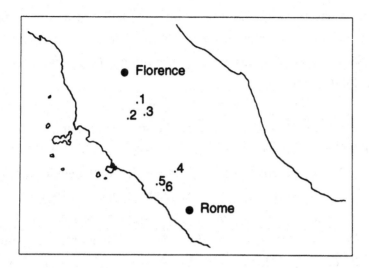

Fig. 1. Geographical localization of the six studied gas springs in Italy: 1 Vagliagli (VA), 2 Carpineto (CV), 3 Castiglioni (CA), 4 Solfatara (SV), 5 Canale Monterano (CB), 6 Caldara di Manziana (CM).

containing only one species were not processed; to avoid information noise, the original matrix was reduced by excluding species occurring in only one relevè (Gauch, 1982). A Minimum Variance clustering method was applied to a dissimilarity matrix of Euclidean Distance after data standardization by range for each variable. Principal Component Analysis (PCA) was performed on the correlation matrix. The relevès table was then ordered according to the multivariate analysis results.

After the floristic sampling, four 1m²-quadrats, two replicates in the two physiognomically different aspects of the grass community as recognized in the field, were established and harvested to compare some structural and vitality-fertility parameters of the dominant species (Zoller, 1954): height of the leaf canopy, number and diameter of tussocks, number of flowering stems and above-ground biomass (dry matter) were determined.

In a third step, four soil profiles were dug in the area of the herbaceous vegetation and four samples from the rhizosphere layer (at a depth of 5-15 cm) were collected and dried: pH, available phosphorus, exchangeable potassium, total nitrogen, cation exchange capacity and total organic matter were measured. Soluble iron, copper, nickel, zinc and exchangeable aluminium, magnesium, calcium and sodium were also measured on soil samples taken at SV. Metal ions were determined by means of the EDTA extractive solution method.

Finally, at all the sites one-hour measurements of CO_2 atmospheric concentration were taken in the area of the grass community by means of an Infrared Gas Analyser (PPS EGM-1, Hitchin, UK) to evaluate the degree of CO_2 enrichment at the canopy level. More detailed analysis of atmospheric CO_2 concentration gradients under different weather conditions were already reported for SV (Miglietta & Raschi, 1993) and for CA, CM, CV (Miglietta *et al.,* 1993a). Taxonomic nomenclature follows Tutin *et al.,* (eds.), 1964-1980 - *Flora Europaea*, Vols. 1-5.

RESULTS

Composition

The herbaceous vegetation was developed closely around the gas vents of all the examined sites, occupying most of the flat areas, the hollows without permanent water bodies and the low banks of the gas springs. The only discontinuities of this vegetation were fragmentary but often large "aphitoic" areas which separated the grassland from the main vents in the larger springs. The grassland reached its maximum extent (about 90 m in diameter) in the large, flat springs (CM, SV), whereas in the narrower ones with steep banks (CA, CV) it formed a ring not more than 8 m in diameter.

The relevès showed that the grass community had a rather consistent and homogeneous species composition in all the examined sites (Table 1). It was formed by very few perennial grasses and sedges and largely dominated by dense stands of *Agrostis canina* subsp. *monteluccii*, an ecological endemism of likely polytopical origin recently described at the subspecific level (Selvi, 1994). In eight relevès this grass was found to constitute dense monospecific stands with 70-100% of ground cover, an uncommon feature for a species without genetically fixed systems of vegetative propagation such as the subsp. *monteluccii*.

TABLE 1 Ordered table of the *Agrostidetum caninae* subsp. *monteluccii*; type relevè: no 24; other monospecific samples of the association (nos. 25-31) are omitted. (*) S sandstone; L clay; I ignimbrite; T tufo.

	dry aspect														wet aspect									
Site	CM	SV	VA	SV	SV	CM	CM	CB	CM	SV	SV	VA	SV	CM	CM	CM	CM	CV	CB	VA	CV	CB	CA	CA
N° sample	24	9	13	14	10	12	11	16	15	21	17	18	19	20	1	2	3	4	5	6	7	8	22	23
Surface sqm	20	15	15	15	20	20	18	16	20	20	12	14	15	16	20	20	20	16	15	12	15	18	16	12
Parent rock (*)	I	L	S	L	L	I	I	T	I	L	L	S	L	I	I	I	I	S	T	S	S	T	S	S
Altitude (m)	260	270	460	270	270	260	260	360	260	270	270	470	270	260	260	260	260	500	360	470	500	360	320	320
Aspect					W			S		SW	E	SE	NW											
Slope (°)					10			5		5	8	2	15											
N° species	1	2	2	2	3	3	3	3	4	3	2	2	2	5	3	4	3	3	3	2	3	3	3	2
Character-species:																								
Agrostis canina ssp. monteluccii	5	5	5	5	4	4	4	4	4	4	4	5	5	3	3	3	4	3	3	2	4	3	3	3
diff. species of the dry aspect:																								
Holcus lanatus		+	1	+	1	+	+	+	1															
Carex hirta					+	+			+		1			+										
others:																								
Poa annua																		1						
Poa trivialis										+		+	+						+		1	1		
Deschampsia flexuosa									+	+											+			
Betula pendula														+								1		
Pteridium aquilinum							1							+										
Rumex acetosella								+						+										
diff. species of the wet aspect:																								
(Molinio-Juncetea Br.Bl. 1947)																								
Juncus articulatus															1	1	2	2	+	1				
Molinia coerulea ssp. arundinacea																							1	1
Scirpus holoschoenus																							1	
others:																								
Scirpus lacustris ssp. tabaernemontani															1	+	1							
Juncus effusus																2								

Considering the distance between the studied sites and their localization in different phytoclimatic areas, the constant presence of this vegetation type and its floristical homogeneity was a surprising fact. In a phytosociological sense *A. monteluccii* could therefore be considered as the character and dominant species of an association endemic to CO_2 springs, the *Agrostidetum caninae* subsp. *monteluccii*. Although in its typical form this association was a pure stand of its character-species, both cluster (Fig. 2) and principal component analysis (Fig. 3) produced a consistent discrimination between two floristical groups of relevès.

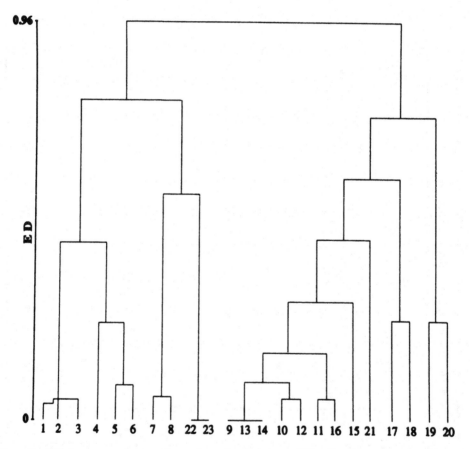

FIG. 2. Minimum variance clustering dendrogram (ED Euclidean Distance) of phytosociological sample plots.

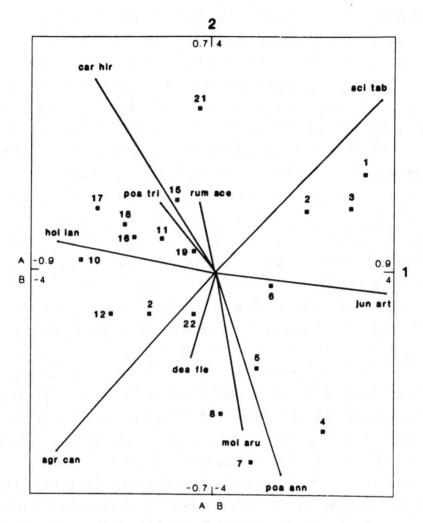

FIG. 3. Principal Component Analysis biplot of sample plots (B scale) and species (A scale). First (horizontal) component eigenvalue = 27 %, second (vertical) component eigenvalue = 15 %. Plots 13, 14, 20, 23 overlap respectively to 9, 9, 19, 22.

The group including plots n° 1, 2, 3, 4, 5, 6, 7, 8, 22, 23, which were largely gathered on the positive part of the first PCA component, represented a wet aspect of the typical association.

In depressions and hollows subjected to winter waterlogging and seasonal flooding as in those closely around the gas vents, the Agrostidetum often came into contact with an helophytic community of standing water bodies, dominated by the subalophytic sedge *Scirpus lacustris* subsp. *tabaer-nemontani*.

In these ecotonal areas *A. monteluccii* became much more sparse (not more than 60% of ground cover) and mixed with some wetland species of the syntaxa of the Molinio-Juncetea Br.Bl. 1947 as *Juncus articulatus*, *Molinia caerulea* ssp. *arundinacea* and *Scirpus holoschoenus*, as well as sparse *Scirpus tabaernemontani* itself. The second group was a dry aspect of the typical association, which included plots n° 9, 13, 14, 10, 12, 11, 16, 15, 21, 17, 18, 19, 20, gathered on the negative part of the first PCA component. This community type was generally localized on soils subjected to a slight summer drought, which occurred in the peripheral areas of the gas springs. At the Latial sites, this aspect of the association came into contact, in its more peripheral areas, with the acidophilic shrublands belonging to the association Adenocarpo complicati-Cytisetum scoparii Blasi *et al.*, 1991 recently described for the volcanic area of Monti Cimini in northern Latium (Blasi *et al.*, 1991). In the Tuscan springs the shrub communities which externally delimited the Agrostidetum area had the same acidophilic character but they were dominated by *Calluna vulgaris*, *Erica arborea* and *Cytisus scoparius*, with some other taxa of the order Cytisetalia scopario-striati Riv.Mart. 1974 and of the alliance Pruno-Rubion ulmifolii De Bolos 1954. The most frequent species of the dry aspect of the Agrostidetum were the two acidocline grasses *Holcus lanatus* and *Carex hirta*; other plants observed here, all with few scattered individuals and a low percentage of ground cover, were mostly grasses such as *Poa annua*, *Poa trivialis* and *Deschampsia flexuosa*. However, the sharp discrimination produced by the two PCA axes was more explainable in terms of a parallel variation of a complex gradient involving at least two other important factors (soil acidity and CO_2 atmosperic concentration) rather than that of a simple soil moisture gradient.

Ecology

Soil profiles of the typical Agrostidetum always showed a layer, at least 25 cm deep, made up of dark peat-like material over clays or materials of volcanic origin (CM, SV, CB) or on siliceous sandstones (VA, CV, CA). Chemical analysis of the peaty substratum in which *Agrostis* was rooted (Table 2) showed pH values ranging from 2.4 to 3.7. The only

TABLE 2. Chemical parameters of soil samples collected at each gas spring in monospecific stands of the *Agrostidetum caninae* subsp. *monteluccii.*

Site	CA	VA	CV	SV	CM	CB
pH (1:2.5)	3.3	3.3	3.1	2.4	2.6	3.7
Available phosphorus, P_2O_5 (ppm)	23.9	98.4	1082	38	52	211.5
Exchangeable potassium, K_2O (mg/100g)	12.8	9.5	16	10.9	37.4	117.4
Total nitrogen (%)	0.39	0.22	0.51	0.36	0.24	0.25
Organic matter (%)	6.08	3.2	7.78	11.1	4	3.46
Cation exch. capacity (meq/100g)	9.9	5.7	16.9	33	33	13
Soluble iron (ppm)	-	-	-	3350	-	-
Soluble copper (ppm)	-	-	-	2	-	-
Soluble nickel (ppm)	-	-	-	1.5	-	-
Soluble zinc (ppm)	-	-	-	4.8	-	-
Soluble aluminium (ppm)	-	-	-	400	-	-
Exchangeable odium (meq/100 g)	-	-	-	0.43	-	-
Exchangeable calcium (meq/100g)	-	-	-	0.54	-	-
Exchangeable magnesium (meq/100 g)	-	-	-	0.23	-	-

microbiological activity in these strongly acid soils is due to species-poor communities of acid-tolerant *Chlorophyceae*, *Euglenophyceae* and *Chrysophyceae* (Albertano, 1993; Pinto, 1993), implying therefore a very slow cycle of decomposition of the organic matter and consequently the production of thick and poorly oxygenated peat layers. Soil hyperacidity is often a toxic factor *per se*, but it is also the cause of several anomalies in the balance of nutrients, especially of phosphorus (Hewitt, 1952; Foy & Brown, 1964). However, the most selective factor for most species is the increased solubility of plant toxic cations such as Al^{3+} (Hewitt, 1952; Kennedy, 1986), which in fact reached 400 ppm at SV, a concentration well above the toxicity thresholds outlined by Clarkson (1966) within the Al-tolerant genus *Agrostis*.

The atmospheric CO_2 concentrations in the area of the monospecific Agrostidetum taken at canopy height ranged between 1750 and 600 μmol mol^{-1} at SV, 10.000 and 750 μmol mol^{-1} at CA and 600-400 μmol mol^{-1} at CM, CB, CV and VA. As can be seen a significant atmospheric CO_2-

enrichment at the canopy height was always present, despite wide fluctuations due to air turbulence.

The "fertilization effect" of elevated atmospheric CO_2 and the possibility of a long-term directional selection of this factor on natural plant communities are among the most important topics of the current ecological research (Idso, 1989; Bazzaz, 1990; Patterson & Flint, 1990; Woodward et al., 1991; Bazzaz & Mc Connaughay, 1992; Rochefort & Woodward, 1992; Johnson, Polley & Mayeux, 1993; Possingham, 1993; Körner & Miglietta, 1994). According to Woodward et al., (1991), if genotypes do differ in CO_2 responses within a species, then natural selection may operate to change the current genetic balance of the species as levels of atmospheric CO_2 continue to rise. A. monteluccii might provide a valuable key to test this hypothesis, since individuals of this plant raised from seeds of populations of the gas springs showed a decreased biochemical capacity for CO_2 assimilation, a decreased stomatal conductance and a decreased RuBP activity, which were preliminarily considered as a genetically-controlled expression of a microevolutionary adaptation to elevated CO_2 environments (Fordham et al., 1996, in this volume).

Structure

At each gas spring two 1 m² quadrats ("N") were laid inside representative stands of the monospecific Agrostidetum and two ("F") in representative stands of its "dry aspect" as physiognomically recognized in the field. The height of the leaf canopy in "N" quadrats was higher than that in "F" ones at all sites (Fig. 4A) and the mean difference between them was 16 cm; however in both quadrats the vegetation was much taller than that usually reported for A. canina in mixed grass communities (Grime, Hodgson & Hunt, 1988). At CM and SV the Agrostidetum canopy reached the remarkable height of 1 m. Tussocks of A. monteluccii were on average 10 cm larger in "N" quadrats, but their number was higher in "F" quadrats at all sites (Fig. 4B). The above-ground biomass was always higher in "N" quadrats (Fig. 4C), their mean value being 489 g_{dw} m⁻² compared to 309 g_{dw} m⁻² in "F" quadrats. The mean number of flowering stems was higher in "N" quadrats than in "F" quadrats, but at three sites (VA, SV, CB) this number was higher in "F" quadrats (Fig. 4D). No clear trend of variation was found in the number of flowering stems between the two community types, also in consideration of the wide variability observed in this parameter (CV = 42% and 49%).

However, in several species, fertility and vitality parameters are known to be often inversely correlated (Barkman, 1989).

The above structural variations accounted for the sharp difference in physiognomy between the two community types: the typical Agrostidetum was formed by dense, large turfs with a tall canopy and often 100% cover, whereas the structure of the dry aspect with *Holcus* and *Carex* was characterized by more numerous but smaller tussocks and by a lower density of the canopy.

The outstanding vegetative vigour of *A. monteluccii* in the typical association was likely to be connected to the absence of soil drought in summer and possibly to a higher atmospheric CO_2 average concentration throughout the year.

At most of the gas springs the CO_2 concentration was in fact found to decrease with the distance from the gas vents, along a horizontal gradient which was dependent on the surface of the sites and on their morphology (Miglietta *et al.*, 1992; Miglietta & Raschi 1993; Miglietta *et al.*, 1993a). Along with this factor, the lower water availability provides the key to explain the decreased competitive capacity of *A.canina* in the peripheral areas of the springs and therefore the higher species richness of this community.

DISCUSSION

The only known reports on the vegetation of acid geothermal gas springs were those by the first author (Montelucci 1947, 1949, 1977) who noticed and described the populations of *A. monteluccii* around some CO_2 springs in Latium. Owing to the rarity of *A.canina* in lowlands of peninsular Italy, this author considered the isolated stations as relicts of the Würmian glacial age strictly localized on the

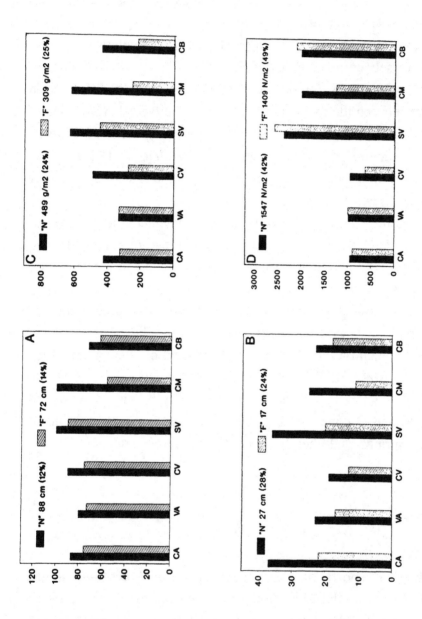

FIG. 4. Structural, vitality and fertility parameters of *A. canina* subsp. *monteluccii* measured in representative stands of the typical Agrostidetum ("N"), and in its dry aspect with *Holcus lanatus* and *Carex hirta* ("F"). Each measurement is the mean of 2 replicates. For both "N" and "F" quadrats the mean values and the coefficients of variation (in brackets) are given. On the X axis are the abbreviations of the sites (see text). A) height (cm) of the leaf canopy; B) diameter (cm) and number (N°m², above the bars) of tussocks; C) above-ground biomass (g$_{dw}$ m^{-2}); D) number of flowering stems (N°m^{-2}).

acid peat surrounding the vents. An interesting and similar situation was described by Von Faber (1925) in the solfataras of Java, in which pioneer paucispecific plant communities with *Ericaceae* were able to colonize the acid and hot substrates with a high content of soluble aluminium.

The species poorness of the Agrostidetum is a consequence of the overwhelming competitive advantage that its dominant has over other species, due to its capacity to tolerate strong soil acidity and high Al^{3+} content and, possibly, to exploit the elevated atmospheric CO_2 concentration. As a consequence of the selective pressure of these environmental constraints, the typical Agrostidetum is not a "true association" of different plants but represents a typical case of ecologically specialized "stress tolerator" (*sensu* Grime, 1979) monospecific vegetation. Consequently, the synecology of this association is the autoecology of its characteristic and dominant species.

The ecological specialization of *A. canina* subsp. *monteluccii* provides further evidence of the capacity known in the genus *Agrostis* to rapidly select populations able to colonize and survive in hostile environments such as lead and zinc poisoned soils (Bradshaw 1952, 1960; Wu, Bradshaw & Thurman, 1975). This is the case of the rapid development of zinc tolerant populations of *A. canina* growing in the vicinity of galvanized fences, as reported by Bradshaw, McNeilly & Gregory (1965). This evolutionary capacity was explained by the presence of genetic variability for tolerance at very low frequency in normal populations on which selection can act with considerable rapidity (Walley, Khan & Bradshaw, 1974).

It was recently supposed (Selvi, 1994) that the presence of supranumerary B-chromosomes in many *Agrostis* species (Björkman, 1951, 1954), as in *A. monteluccii* itself (Fiorini *et al.*, 1992), could be a key factor of microevolutionary advantages at the population level, since it is known that B-chromosomes induce an increase in available sites for chiasma formation during meiosis (Rees, 1974; Vosa, 1983) and therefore an increased genetic variability. Indeed, this type of directional selection could explain the clear polytopic origin of *A. canina* subsp. *monteluccii* (Selvi, 1994), as a consequence of the geographical distance among the populations which has probably prevented genetic exchanges and allowed an independent and parallel microevolution from a common parent genetic pool.

"Fragments" of the Agrostidetum are also known for other mineral gas springs of Tuscany, Latium and Campania (Miglietta *et al.*, 1995), and it is probably present in other still unknown gas springs of central-western Italy. The presence of this association and its homogeneity in rather different phytoclimatic areas of peninsular Italy are evidence of its azonal character, in the sense that it is linked to local and peculiar edaphic (and atmospheric?) conditions rather than to macroclimate.

For all these reasons the syntaxonomical placing of the Agrostidetum can not be immediate, given that no syntaxa of strongly acidophilic vegetation of geothermic sites are currently available in the known European literature. The evaluation of the degree of floristical and ecological similarity with other *A. canina*-dominated communities provides only weak indications.

According to some authors *A. canina* does not have a clear phytosociological characterization, being considered mostly a gap-colonizing species rather than a dominant in species-rich grass communities (Grime *et al.*, 1988). Other European authors, however (Gadeceau, 1909; Gaume, 1924; Wiegleb, 1977) recognized some community types dominated by *A. canina*, mostly in Atlantic areas of north-western Europe. Two of these, the *Agrostidetum caninae euatlanticum* (Gadeceau, 1909; Lemeè, 1938) and the *Agrostis canina* community (Gaume, 1924) were reported to form a ring on acidic oligotrophic soils surrounding ponds in lowlands of Atlantic France and Spain (Gaume, 1924; Duvigneaud, 1949). These associations were placed by Duvigneaud (1949) within the order *Molinio-Caricetalia fuscae* Duv. 1949; indeed, for its acidophilic character, *A. canina* is considered a character-species of the association *Caricetum fuscae* Br.Bl. 1915 (Braun Blanquet, 1971), of the alliance *Caricion canescentis-goodenoughii* Nordh. 1937 (Vanden Berghen, 1952) and of the order *Caricetalia fuscae* Koch 1926 em. Br.Bl. 1949 (Steiner, 1993). However, all the above syntaxa describe community types which are spread over Atlantic and subcontinental areas of Europe and restricted to the (sub)alpine belt in the Mediterranean (Dierssen, 1978). Consequently it can be said that even though the oligotrophic species-poor communities of European acidic fens and mires have more ecological affinities than any other with the Agrostidetum, the syntaxonomical placing is still premature and requires further coenological investigations.

Further studies on plant communities of other acidic gas vent areas over a wide geographical scale could lead to the circumscription of syntaxa of higher rank, recognizing the floristical and ecological autonomy of these still poorly-known community types.

ACKNOWLEDGEMENTS

I wish to acknowledge Dr. I. Bettarini for her help in field work and Dr. F. Miglietta and Dr. A. Raschi for their encouragement and suggestions. The cost of the research has been partially covered by EU, Programme Environment, contract EV5V-CT93-0432.

REFERENCES

Albertano, P. (1993) Microalgae from sulphuric acid environments - In *Algae, man and environments* (ed. W. Wiessner). Chapman and Hall, London.

Baldi, P., Decandia, F., Lazzarotto, A. & Calamai, A. (1974) Hydrochemistry of the region between Monte Amiata and Rome. *Geothermics*, 2, 124-141.

Barkman, J.J. (1989) Fidelity and character-species, a critical evaluation. *Vegetatio*, 85, 105-116.

Bazzaz, F.A. (1990) The response of natural ecosystems to the rising global CO_2 levels. *Ann. Rev. Ecol.& Syst.*, 21, 167-414.

Bazzaz, F.A. & McConnaughay, K.D.M. (1992) Plant-plant interactions in elevated CO_2 environments. *Australian Journal of Botany*, 40, 547-563.

Björkman, S.O. (1951) Chromosome studies in *Agrostis* (a preliminary report). *Hereditas*, 37, 465-468.

Björkman, S.O. (1954) Chromosome studies in *Agrostis*. II. *Hereditas*, 40, 254-258.

Blasi, C., Cavaliere A., Abbate G. & Scoppola A. (1991) I cespuglieti del comprensorio vulcanico Cimino-Vicano (Lazio, Italia Centrale). *Ann. Bot. (Roma)*, 48(7), 1-15.

Bradshaw, A.D. (1952) Populations of *Agrostis* tenuis resistant to lead and zinc poisoning. *Nature*, 169, 1098.

Bradshaw, A.D. (1960) Population differentiation in *Agrostis* tenuis Sibth. II. Populations in varied environments. *New Phytologist*, **59**, 92-103.

Bradshaw, A.D., McNeilly, T.S. & Gregory, R.P.G. (1965) Industrialization, evolution and the development of heavy metal tolerance in plants. In *Ecology and the industrial society* (eds. G.T. Goodman *et al.*), pp. 327-343. Oxford.

Braun Blanquet, J. (1971) Übersicht der Pflanzengesellschaften der rätischen Alpen im Rahmen ihrer Gesamtverbreitung. Teil III. Flachmoorgesellschaften (Scheuchzerio-Caricetea fuscae), Veröff. Geobot. Inst. ETH Stift. Rübel, Zürich.

Clarkson, D.T. (1966) Aluminium tolerance in species within the genus *Agrostis*. *Journal of Ecology*, **54**, 167-178.

Dierssen, K. (1978) Some aspects of the classification of oligotrophic and mesotrophic mire communities in Europe. In *La végétation de sols tourbeux* (ed. J.M. Gehu), *Coll. Phytosoc.*, 7, 399-423.

Duchi, V., Minissale, A. & Romani, L. (1985) Studio geochimico su acque e gas dell'area geotermica lago di Vico-M. Cimini (Viterbo). *Atti Soc. Tosc. Sci. Nat., Mem., Ser.* A **92**, 237-254.

Duvigneaud, P. (1949) Classification phytosociologique des tourbières de l'Europe. *Bull. Soc. Roy. Bot. Belg.*, **81**, 58-129.

Fiorini, G., Miglietta, F., Raschi, A., & Selvi, F. (1992) Analisi cariologica di alcune popolazioni di *Agrostis canina* L. (Gramineae) dell'Italia centro-occidentale. *Informatore Botanico Italiano*, **24**(1-2), 32-38.

Fordham, M., Barnes, J., Broadmeadow, M., Raschi, A., Bettarini, I., Miglietta, F. (1994) Evidence for micro-evolutionary adaptation to elevated CO_2 in *Agrostis canina* (Abstract). In: International Conference "The Manipulation of Photosynthetic Carbon Metabolism in Crop Improvement", Rothamsted Experimental Station, Harpenden, Herts, UK, 6-9 Oct. 1994.

Foy, C.D. & Brown, J.C. (1964) Toxic factors in acid soils. II Differential aluminum tolerance of plant species. *Soil Sci.*, Soc. Am. Proc., **28**, 27.

Gadeceau, E. (1909): Le lac de Grand-Lieu, p. 115. Paris.

Gauch, H.G. (1982) *Multivariate analysis in community ecology*. Cambridge Studies in Ecology. Cambridge University Press, Cambridge.

Gaume, R. (1924) Les associations végétales de la forêt de Preuilly (Indre-et-Loire). *Bull. Soc. Bot. Fr.*, **71**, 158-171.

Grime, J.P. (1979) *Plant strategies and vegetation processes*. John Wiley.

Grime, J.P., Hodgson, J.G. & Hunt, R. (1988) *Comparative plant ecology*. Unwin Hyman Press, London.

Hewitt, L.J. (1952) A biological approach to the problems of soil acidity. *Trans. Int. Soc. Soil Sci.*, Jt. Meet. Dublin, **1**, 107-118.

Idso, S.B. (1989) *Carbon dioxide and global change: Earth in transition*. IBR Press, Temple.

Johnson, H.B., Polley, H.W. & Mayeux, H.S. (1993) Increasing CO_2 and plant-plant interactions: effects on natural vegetation. *Vegetatio,* **104/105,** 157-170.

Kennedy, I.R. (1986) *Acid soils and acid rain,* pp. 167-174. Research Study Press, England.

Koch, G.W. (1994) The use of natural situations of CO_2 enrichment in studies of vegetation responses to increasing atmospheric CO_2. In *Design and Execution of Ecosystem Experiment* (eds. E.D. Schulze, & H.A. Mooney), Ecosystem Research Report n.6, Commission of the European Communities.

Körner, C. & Miglietta, F. (1994) Long-term response of mediterranean grassland and forest trees to elevated CO_2 concentrations. *Oecologia,* **99,** 343-351.

Miglietta, F. & Raschi, A. (1993) Studying the effect of elevated CO_2 in the open in a naturally enriched environment in Central Italy. *Vegetatio* **104/105,** 391-400.

Miglietta, F., Raschi, A., Bettarini, I. & Selvi, F. (1992) Vegetazione di aree arricchite da emissioni naturali di anidride carbonica dell'Italia centrale. *Acqua Aria,* **9,** 869-874.

Miglietta, F., Raschi, A., Bettarini, I., Resti, R. & Selvi, F. (1993a) Natural CO_2 springs in Italy: a resource for examining long-term responses of vegetation to rising atmospheric CO_2 concentrations. *Plant, Cell and Environment,* **16,** 873-878.

Miglietta, F., Raschi, A., Resti, R. & Badiani, M. (1993b) Growth and onto-morphogenesis of soybean in an open naturally CO_2 enriched environment. *Plant, Cell and Environment,* **16,** 909-918.

Miglietta, F., Badiani, M., Bettarini, I., Van Gardingen, P., Selvi, F., Raschi, A. (1995) Preliminary studies of the long-term response of Mediterranean vegetation around natural CO_2 vents. In *Anticipated effects of a changing global environment in Mediterranean-type ecosystems* (eds. J.M Moreno & W.C. Oechel), pp. 102-120.

Montelucci, G. (1947) Investigazioni botaniche nel Lazio. 4. aspetti della vegetazione dei travertini alle Acque Albule (Tivoli). *N. Giorn. Bot. Ital.,* **54,** 494-504.

Montelucci, G. (1949) Alcune piante notevoli della vegetazione laziale. *N. Giorn. Bot. Ital.,* **56,** 366-418.

Montelucci, G. (1977) Note preliminari sulla flora e sulla vegetazione delle cerrete di Manziana e di Canale Monterano - In: Ricerche ecologiche, floristiche e faunistiche nel comprensorio Tolfetano-Cerite-Manziate, quad. **27,** 51-73. Accademia Naz. dei Lincei, Roma.

Panichi, C. & Tongiorgi E. (1975) Carbon isotopic composition of CO_2 from springs, fumaroles, mofettes and travertines of Central and Southern Italy: a preliminary prospection method of geothermal area. In: *Proceedings of the second United Nations symposium on the development and use of geothermal resources*, San Francisco, CA, USA, Vol. 1 pp. 815-825.

Patterson, D.T. & Flint, E.P. (1990) Impact of carbon dioxide trace gases and climate change on global agriculture. *ASA Special Publication*, **53**, 83-110.

Pinto, G. (1993): Acid-tolerant and acidophilic algae from Italian environments. *Giornale Botanico Italiano*, **127** (3), 400-406.

Podani, J. (1991) SYNTAX IV. Computer programs for data analysis in ecology and systematics. In *Computer assisted vegetation analysis* (eds. E. Feoli & Orlóci), pp. 437-452. Kluwer Acad. Publ., Netherlands.

Possingham, H.P. (1993) Impact of elevated atmospheric CO_2 on biodiversity: mechanistic population-dynamic perspective. *Australian Journal of Botany*, **41**, 11-21.

Rees, H. (1974) B-Chromosomes. *Sci. Progr. Oxford*, **61**, 535-554.

Rochefort, L. & Woodward, F.I. (1992) Effects of climate change and a doubling of CO_2 on vegetation diversity. *Journal of Experimental Botany,* **43** (253), 1169-1180.

Selvi, F. (1994) *Agrostis canina* L. subsp. monteluccii Selvi. *Webbia*, **49**, 51-58.

Sollevanti, F. (1983) Geologic, volcanologic and tectonic setting of the Vico-Cimini area, Italy. *Jour. Volc. Geoth. Res.*, 17, 203-217.

Steiner, G.M. (1993) Scheuchzerio-Caricetea fuscae. In *Die Pflanzengesellschaften Österreichs* (eds. G. Grabherr & L. Mucina). Teil III, Natürliche waldfreie Vegetation, G. Fischer Verlag Jena, Stuttgart, New York.

Touring Club Italiano (1957) L'Italia fisica, pp. 117-123, Milano.

Vanden Berghen, C. (1952) Contribution a l'étude des bas-marais de Belgique. *Bull. Jard. Bot. Ét. Bruxelles*, **22** (1,2), 1-63.

Von Faber, F. (1925) Untersuchungen über die Physiologie der javanischen Solfataren-Pflanzen. *Flora*, **118/119**, 89-110.

Vosa, C.G. (1983) The ecology of B-chromosomes in *Listera ovata* (Orchidaceae). *Caryologia*, 36(2), 113-120.

Walley, K.A., Khan, M.S.I. & Bradshaw, A.D. (1974) The potential for evolution in heavy metal tolerance in plants. I. Copper and zinc tolerance in *Agrostis tenuis*. *Heredity*, **32**, 309-319.

Westhoff, V. & Van der Maarel, E. (1978) The Braun Blanquet approach. In *Classification of plant communities* (ed. R.H. Whittaker), pp. 638-640. Junk, The Hague.

Wiegleb, G. (1977) Vegetation und Umweltbedingungen der Oberharzer Stauteiche heute und in Zukunft - Mskr.

Wisheu, I.C. & Keddy, P.A. (1992) Competition and centrifugal organization of plant communities: theory and tests. *J. Veg. Sci.*, **3**, 147-156.

Woodward, F.I., Thompson, G.B. & McKee, I.F. (1991) The effects of elevated concentrations of carbon dioxide on individual plants, populations, communities and ecosystems. *Annals of Botany*, **67** (suppl.1), 23-38.

Wu, L., Bradshaw, A.D. & Thurman, D.A. (1975) The potential for evolution of heavy metal tolerance in plants. III. The rapid evolution of copper tolerance in *Agrostis stolonifera. Heredity*, **34**, 165-187.

Zoller, H. (1954) Die Typen der Bromus erectus. Wiesen des Schweizer Juras. Ihre Abhängigkeit von den Standortsbedingungen und wirtschaftlichen Einflüssen und ihre Beziehungen zur ursprünglichen Vegetation. *Beitr. Geobot. Landesaufn. Schweiz.* **33**, 1-309.

Studying morpho-physiological responses of *Scirpus lacustris* from naturally CO₂-enriched environments

I. BETTARINI, F. MIGLIETTA AND A. RASCHI

I.A.T.A. - C.N.R., P.le delle Cascine 18 - 50144 Firenze, Italy.

SUMMARY

Gas vent areas, where CO_2 of deep origin is naturally released into the atmosphere, provide a valuable opportunity to study long-term effects of CO_2 enrichment on natural vegetation. A study of adaptive traits in relation to elevated CO_2 was carried out on genetically isolated populations of *Scirpus lacustris*, a rhizomatous emergent wetland sedge growing at several CO_2 springs in Italy. Plants were grown in the laboratory from rhizomes collected in CO_2 springs and control sites and their photosynthetic capacities were compared using gas-exchange techniques. Carbon isotope discrimination, nitrogen content and stomatal density were then measured in the field on plants of the same species growing along a transect traced from gas vents outwards. Photosynthetic rates of plants grown in the laboratory from CO_2 springs were not significantly different from those of control plants. The absence of decline in photosynthetic capacity with increasing external CO_2 concentration was supported by results from plants grown along the CO_2 gradient in the field. They showed no change in nitrogen content and in Ci/Ca ratio, and exhibited a downward regulation of stomatal density. The possible role of wetland ecosystems as sinks for atmospheric CO_2 under greenhouse climate conditions is discussed.

INTRODUCTION

Short-term exposure to elevated CO_2 concentration is known to cause an increase in the photosynthetic rate in most C_3 species, whereas long-term growth under elevated CO_2 concentration often results in a downward regulation of photosynthetic capacity (Sage, Sharkey & Seemann, 1989;

Arp, 1991). The critical question during the next century is whether the terrestrial biosphere with respect to atmospheric CO_2, will provide a sink, moderating the rate of global change, or an additional source of CO_2 enhancing greenhouse warming. Little is known about adaptation of plants to a chronic elevation of atmospheric CO_2 concentration and in particular about the biochemical capacity for CO_2 fixation or stomatal functioning.

Current research is increasingly focusing on the response of plants grown in the plumes of natural CO_2 sources, primarily springs and caves (Sage, 1994). Previous data provided contrasting evidence on possible adaptive traits. A population of *Boehmeria cylindrica* collected in the proximity of a CO_2 spring in Florida showed enhanced growth rates as compared to a control population (Woodward, Thompson & Mc Kee, 1991) and the tissue concentration of total non-structural carbohydrates was 38-47 % higher in species from a grassland community growing in a CO_2 spring in central Italy than in plants growing outside the CO_2-enriched area (Körner & Miglietta, 1994).

Nitrogen limitation seems to affect the photosynthetic capacity of leaves of *Erica arborea* growing in a CO_2 spring in southern Italy, and to interact with stomatal conductance in regulating isotopic discrimination (Bettarini *et al.*, 1995a). In contrast preliminary results with the perennial *Ludwigia uraguayensis*, an aquatic herb growing near CO_2 springs in California, showed that there is little difference in the A/Ci response in plants growing under normal atmospheric air with respect to plants growing in the plumes where CO_2 levels were two to four times the ambient (Koch, 1994).

Plants growing in these gas-vent areas may be ecotypically adapted to elevated CO_2 compared to nearby populations of the same species growing outside the enriched areas (Koch, 1994; Miglietta *et al.*, 1993; Miglietta *et al.*, 1994), as found with the wetland sedge *Scirpus lacustris*, which grows at several CO_2 springs in central Italy. It was suggested that given the long-term exposure to elevated CO_2, these populations could have undergone a microevolutionary adaptation to high levels of CO_2, by developing acclimation traits (Amthor, 1995).

The aim of this paper is to investigate the existence of adaptive traits to elevated CO_2 in populations of *Scirpus lacustris*, by comparing the photosynthetic capacity of shoots grown from rhizomes collected in CO_2

springs and in areas with ambient CO$_2$ concentration. The response of net carbon assimilation to intercellular CO$_2$ concentration (A/Ci curves) allows studies of acclimation in terms of biochemical reactions and stomatal processes controlling the long-term response of photosynthetic capacity to CO$_2$ enrichment (Von Caemmerer & Farquhar, 1981; Farquhar & Sharkey, 1982).

Stomatal response of *Scirpus* in the wild, in terms of stomatal density and specific leaf area and Ci/Ca ratio, were determined in a second phase of the research on plants growing in a CO$_2$ spring along a CO$_2$ concentration gradient. N content in shoot tissues was taken as an indicator of changes in photosynthetic capacity.

MATERIALS AND METHODS

Photosynthetic measurements

Rhizomes of *Scirpus lacustris*, buried in permanently submerged muds, were collected in March 1992 from three CO$_2$ springs: Solfatara (Viterbo), Caldara di Manziana (Rome) and Gambassi (Siena), at a distance from the CO$_2$ vents where CO$_2$ concentration was twice the ambient. Rhizomes of *S. lacustris* were also collected in three control areas at ambient CO$_2$ concentration: Lake Sibolla (Lucca), Lake Massaciuccoli (Pisa) and Torre di Castello (Siena).

The rhizomes, deprived of both shoots and roots, were placed in pots and grown from March to July in hydroponic culture using a modified Hoagland nutritive solution. Pots were placed in a growth chamber with a 12-h photoperiod at 20°C during light hours and 15°C in the dark. R.H. was 65%. Irradiance at the top of the canopy was 700 µmol m^{-2} s^{-1}. The CO$_2$ concentration inside the growth chamber was controlled by automatic regulation of pure CO$_2$ through a CIM (PP-System, Hitchin, UK) to maintain 900 µmol mol^{-1} 24 h per day.

Gas exchange measurements were taken in July with an open system to evaluate the photosynthetic capacity of the plants in terms of assimilation response (A) to the intercellular CO$_2$ concentration (Ci). A well-stirred 20 cm long cylindrical cuvette was fed by a controlled air flow (Tylan, USA), in which CO_2 was kept constant. Differential concentration measurements of the air entering and leaving the cuvette were taken with two Infra Red Gas Analyzers (ADC, Hoddesdon, UK) for CO_2 and water vapour. Vapour pressure deficit was maintained constant in the cuvette at 1.8 ± 0.1 KPa. The CO_2 concentration of inlet air (Ca) was altered in 5 steps: CO_2-free air, 350 µmol mol^{-1}, 600 µmol mol^{-1}, 1200 µmol mol^{-1} and 2000 µmol mol^{-1}. Assimilation measurements were taken with all variables steady. Algorithms proposed by Ball (1987) were used to calculate gas exchange.

Isotopic analysis

In July 1993 shoots of *Scirpus lacustris* were collected at Solfatara along a CO_2 transect traced from major gas vents (see Fordham et al., in this volume). The isotopic ratio ($\delta^{13}C$) of shoot dry matter was determined at the Dept. of Biology of Utah University, using a dual inlet mass spectrometer (Finnigan Mat, Delta-S, Hamburg) on shoot samples collected 15 cm below the flowering shoots. $\delta^{13}C$ was calculated as follows:

$$\delta^{13}C = (R_{sample} / R_{standard}) * 1000 \qquad [1]$$

where R_{sample} and $R_{standard}$ are the $^{13}C/^{12}C$ ratio of the sample and of a standard (PDB), respectively. Shoot dry matter isotope discrimination (Δ) was then calculated according to the following equation (Farquhar, Ehleringer & Hubick, 1989):

$$\Delta = (\delta_a - \delta_p)/(1 + \delta_p) \qquad [2]$$

where δ_a is $\delta^{13}C$ of source air, calculated by means of carbon isotope ratio of C_4 plants, artificially grown along the transect (Bettarini et al., 1995b; Fordham et al., in this volume), and δ_p is $\delta^{13}C$ of the *Scirpus* samples. From these values Ci/Ca ratio was calculated as follows:

$$Ci/Ca = (\Delta - a)/(b - a) \qquad [3]$$

where a is the fractionation due to the CO_2 diffusion in the air (4.4‰) and b is the fractionation due to carboxylation (27‰) (Farquhar *et al.*, 1989).

Nitrogen content

N content of oven dried plant material was determined by the micro-Kjeldhal method on the same samples of *Scirpus* used for isotopic analysis. This content was expressed on mass basis ($g_{dw.}/g$) and on shoot area basis ($\mu mol\ m^{-2}$), through determination of the specific shoot area.

Stomatal density

Samples for stomata counting were taken on *Scirpus* at the Solfatara CO_2 spring, on shoots 15 cm below the flowering shoots. Stomatal density (number of stomata per unit area, $N°\ mm^{-2}$) was determined on microphotographs of Rodoith impression (Vazzana *et al.*, 1988).

   Stomatal density was also determined on plants grown artificially at elevated H_2S concentration. Rhizomes of *Scirpus* collected at Solfatara in November 1993, deprived of shoots, were placed in pots in two growth chambers with a 12 h photoperiod at 20°C and 15°C during light and dark respectively. Irradiance at the top of the canopy was 700 $\mu mol\ m^{-2}\ s^{-1}$. The H_2S concentration inside one of the growth chambers was set at 1 $\mu mol\ mol^{-1}$, controlled by means of a Smart Display Control connected to a Polytron-H_2S (Dräeger, Lübeck, Germany). At flowering, samples of shoots were cut 10 cm below the flowering shoots and Rodoith impressions were taken to determine stomatal density on microphotographs (Vazzana *et al.*, 1988).

RESULTS

Photosynthetic measurements

Results of gas-exchange measurements for the two groups of populations of *Scirpus* are reported in Fig. 1. As can be seen gas exchange conducted

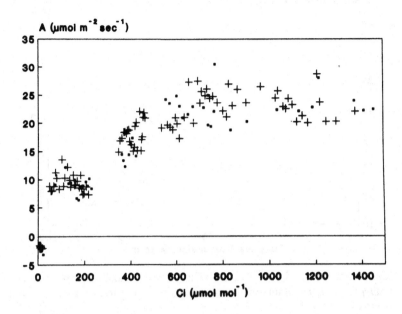

FIG. 1. CO_2 assimilation of *Scirpus lacustris* plotted versus internal carbon dioxide concentration. Legend: (+) plants grown from rhizomes collected in CO_2-enriched areas; (■) plants grown from rhizomes collected in control sites.

Isotopic analysis

Scirpus plants, collected along the CO_2 gradient, showed that the isotopic discrimination (Δ) and Ci/Ca ratio values remained constant with the distance from major vents (Fig. 2). *Scirpus* plants growing at different CO_2 concentrations exhibited similar values in the Ci/Ca ratio (mean value: 0.46±0.02).

Nitrogen content

N content did not statistically vary in plants collected along the transect (mean value: $0.022±0.5*10^{-3}$ $g_{dw.}/g$) (Fig. 3). The constancy of N content was confirmed when expressed on the shoot area basis (mean value: 108.39±4.02 mmol m^{-2}) (Fig. 3).

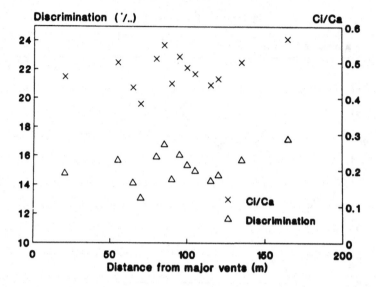

FIG. 2. Isotopic discrimination (Δ) and Ci/Ca (X) ratio calculated solving eq. [4] of shoots of *Scirpus lacustris* plants grown along a directional transect traced from major vents outwards.

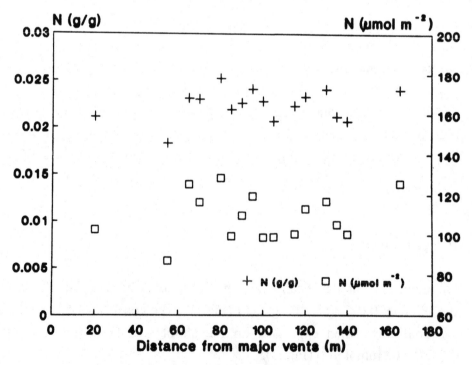

FIG. 3. Shoot N content of *S. lacustris* expressed on a shoot mass (g_{dw}/g) and shoot area (mmol m^{-2}) basis is plotted versus distance from major gas vents.

Stomatal density

Stomatal density in shoots of plants growing at different CO_2 concentrations (Fig. 4) showed a statistically significant decline with decreasing distance from the gas vents.

Values ranged from 108.6 N° mm^{-2} in plants near the gas emissions where CO_2 concentration was about 800 µmol mol^{-1}, to 133.9 N° mm^{-2} in plants growing far along the transect, where CO_2 concentration was about 400 µmol mol^{-1}. The regression line versus the distance (d) from major vents was: s.d.=82.3+0.39*d (R^2=46.6%, cor. coef.=0.68, df=76, P<0.001). Stomatal density of plants of *S. lacustris* artificially grown at elevated H_2S concentrations (mean value: 110.59±2.20 mm^{-2}) was not statistically different from that of plants grown in the control growth chamber (mean value: 109.75±2.66 mm^{-2}).

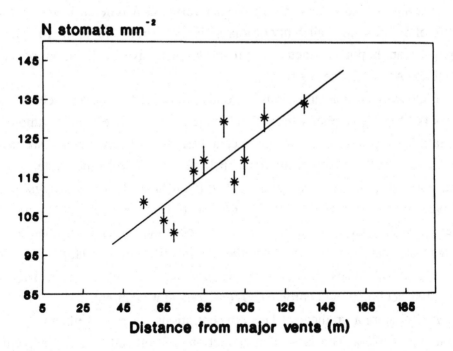

FIG. 4. Stomatal density (N° mm^{-2}) of shoots of *S. lacustris* growing along a CO_2 gradient at Solfatara is plotted versus distance (D) from major gas vents. Slope regression: s.d.=82.3+0.39*D.

DISCUSSION

The absence of decrease in photosynthetic capacity of shoots of *Scirpus lacustris* grown from rhizomes collected both in naturally CO_2-enriched environments and in control sites suggests that, apparently, no evolutionary changes occurred in the overall capacity of photosynthetic CO_2 fixation. This could indicate that there is no evolutionary adjustment in the activity level of Rubisco and in the photosynthetic electron transport capacity. This is in agreement with the idea that the potential evolution of the photosynthetic "machinery" is very low (Morell *et al.*, 1992).

Decrease in photosynthetic capacity after long-term exposure to high levels of CO_2 has been attributed to "end product" inhibition, resulting from an enhanced production of carbohydrates which exceeds the capacity of storage in the sink (Clough, Peet & Kramer, 1981; Mott, 1990; Stitt, 1991). Accumulation of non-structural soluble carbohydrates is a common feature of plants exposed to elevated CO_2 (Körner & Arnone, 1992), even under natural conditions in CO_2 springs (Körner & Miglietta, 1994). In the case of *Scirpus* the availability of a large carbohydrate sink in the rhizomes may be an important factor in preventing negative acclimation of this species (Arp & Drake, 1991).

Elevated CO_2 concentrations were also measured in the aerenchyma of shoots of *Scirpus* plants grown at ambient CO_2 concentration (unpubl. data); this appears to be a common characteristic of many wetland species, being the CO_2 derived from the substrate via the rhizome system and transported in the lacunae by mass flow (Constable, Grace & Longstreth, 1992; Longstreth, 1989). Although no direct evidence for the utilization of aerenchymatic CO_2 exists, the variation occurring under light conditions suggested that there was photosynthetic utilization (Brix, 1988; Constable *et al.*, 1992). These features of *Scirpus* plants could explain why a negative acclimation to chronic CO_2 enrichment was absent.

At Solfatara a gradient of H_2S concentrations occurs together with the gradient of CO_2. The lack of a reduction in stomatal density in plants grown in the laboratory under high levels of H_2S excludes that the presence of this gas could affect stomatal density along the CO_2 gradient.

Increase in concentration of ambient CO_2 provides increase in stomatal resistance; this enhances net CO_2 assimilation and reduction in water loss (Morrison, 1987) because of stomatal closure or because of the reduction in the stomatal pore area per unit surface which increases the exchange resistance between leaf and ambient air (Goudriaan & Unsworth, 1990). There is some evidence that plants can adapt to a changed CO_2 environment through modifications at both morphological and physiological level. Recent studies on dry herbaria specimens (Woodward, 1987; Peñuelas & Matamala, 1990; Paoletti & Gellini, 1993) and from CO_2 springs (Miglietta & Raschi, 1993) support the idea of the occurrence of a decrease in stomatal density under elevated atmospheric CO_2. These permanent changes in the distribution of stomata are likely to be limited to the maximal conductance of leaves (Laisk, 1983; Körner, Scheel & Bauer, 1979).

A decrease in stomatal conductance (gs) associated with high intercellular CO_2 concentration has been considered as a clear example of adaptive response to high CO_2 (Mott, 1990). Since stomatal conductance is variable with both external CO_2 concentration and A, it is difficult to use gs alone to assess changes in the relationship between stomatal behaviour and photosynthesis which may reflect stomatal acclimation to high CO_2 (Sage, 1994). A useful index of possible stomatal acclimation to long-term exposure is the Ci/Ca ratio which assesses the effects of changes in ambient CO_2 on Ci and photosynthesis over a wide range of environmental conditions (Bell, 1982; Ball & Berry, 1982).

On the basis of the leaf photosynthesis model of Von Caemmerer and Farquhar (1981) the observed constant Ci/Ca ratio value with increasing Ca in *Scirpus* plants could be explained by an increase in leaf photosynthetic capacity (A) and a decrease in stomatal conductance (gs). No consistent change in Ci/Ca ratio following long-term CO_2 enrichment was observed in C_3 species from pooled data and in most cases the reduction in g_S maintains Ci/Ca constant across a broad range of Ca (Sage, 1994).

According to the photosynthesis-N relationship (Field & Mooney, 1986; Evans, 1989) the absence of decline in N content both on mass and area basis indicates that photosynthetic capacity was not decreased in *Scirpus* plants growing along the CO_2 gradient concentration. Recent studies showed that leaf N is a good predictor of photosynthetic capacity also for plants grown under elevated CO_2 as for a given Ci, a rise in A was necessarily associated to a rise in leaf N content (Petterson & McDonald, 1992; Nijs, Impens & Van Hecke, 1992).

In conclusion an increase in photosynthetic capacity appears to be ruled out because shoot N concentrations are constant. Therefore a constant Ci/Ca ratio of plants growing across the distance from vents implies increasing stomatal limitation - through fewer stomata and/or changes in stomatal pore length and width.

ACKNOWLEDGEMENTS

The participation of I. Bettarini was during a Ph.D. studentship at the University of Reading funded by the UK NERC TIGER Programme. The authors thank G. Tagliaferri and M. Lanini for their valuable technical contribution. The cost of the research has been partially covered by EU, Programme Environment, EV5V-CT93-0432.

REFERENCES

Amthor, J.S. (1995) Terrestrial higher-plants response to increasing atmospheric [CO_2] in relation to the global carbon cycle. *Global Change Biology*, **1**, 243-274.

Araus, J.L. & Buxo, R. (1993) Changes in carbon isotope discrimination in grain cereals from the North-Western Mediterranean Basin during the past seven millennia. *Australian Journal of Plant Physiology*, **20**, 117-123.

Arp, W.J. (1991) Effects of source-sink on photosynthetic acclimation to elevated CO_2. *Plant, Cell and Environment*, **14**, 869-875.

Arp, W.J. & Drake, B.G. (1991) Increased photosynthetic capacity of *Scirpus olney* after 4 years of exposure to elevated CO_2. *Plant, Cell and Environment*, **14**, 1003-1006.

Ball, J.T. (1987) Calculations related to gas exchange. In *Stomatal function* (eds. E. Zeiger, G.D. Farquhar & I.R. Cowan), pp. 445-476. Stanford Univ. Press, Stanford, California.

Ball, J.T. & Berry, J.A. (1982) The Ci/Ca ratio, a basis for predicting stomatal control of photosynthesis. *Carnegie Institute of Washington Yearbook*, **81**, 88-92.

Bell, C.J. (1982) A model of stomatal control. *Photosynthetica*, **16**, 486-495.

Bettarini, I., Calderoni, G., Miglietta, F., Raschi, A. & Ehleringer, J. (1995a) Photosynthetic capacity and isotopic discrimination of *Erica arborea* L. along a carbon dioxide concentration gradient in a CO_2 spring in Italy. *Tree Physiology*, **15**,321-332.

Bettarini, I., Giuntoli, A., Miglietta, F. & Raschi, A. (1995b) Using stable carbon isotope ratio of C4 plants to monitor mean effective atmospheric CO_2 concentrations. *Agricoltura Mediterranea*, **special volume**, 278-281.

Brix, H. (1988) Light-dependent variations in the composition of the internal atmosphere of *Phragmites australis* (Cav.) Trin ex Steudel. *Aquatic Botany*, **30**, 319-329.

Clough, J.M., Peet, M.M. & Kramer, P.J. (1981) Effects of high atmospheric CO_2 and sink size on rates of photosynthesis of a soybean cultivar. *Plant Physiology*, **67**, 1007-1010.

Constable, J.V.H., Grace, J.B. & Longstreth, D.J. (1992) High carbon dioxide concentrations in aerenchyma of *Thypha latifolia*. *American Journal of Botany*, **79**, 415-418.

Evans, J.R. (1989) Photosynthesis and nitrogen relationship in leaves of C3 plants. *Oecologia*, **78**, 9-19.

Farquhar, G.D. & Sharkey, T.D. (1982) Stomatal conductance and photosynthesis. *Annual Review of Plant Physiology*, **33**, 317-345.

Farquhar, G.D., Ehleringer, J.R. & Hubick, K.T. (1989) Carbon isotope discrimination and photosynthesis. *Annual Review of Plant Physiology and Plant Molecular Biology*, **40**, 503-537.

Field, C. & Mooney, H. (1986) The photosynthesis-nitrogen relationship in wild plants. In *On the economy of plant form and function* (ed. T.J. Givnish), pp. 25-55, Cambridge University Press, Cambridge.

Goudriaan, J. & Unsworth, M.H. (1990) Implications of increasing carbon dioxide and climate change for agricultural productivity and water resources. In *Impact of carbon dioxide, trace gases and climate change on global agriculture* (eds. B.A. Kimball, N.J. Rosenberg & L.H. Allen), pp.111-130, ASA Special Publication n.53.

Koch, G.W. (1994) The use of natural situations of CO_2 enrichment in studies of vegetation responses to increasing atmospheric CO_2. In *Design and execution of ecosystem experiments* (eds. E.D. Schulze & H.A. Mooney), pp. 381-391, Ecosystem Research Report n.6, Commission of the European Communities.

Körner, C., Scheel, A. & Bauer, H (1979) Maximum leaf diffusive conductance in vascular plants. *Photosynthetica*, **13**, 45-82.

Körner, C. & Arnone, J.A. (1992) Response to elevated carbon dioxide in artificial tropical ecosystems. *Science*, **257**, 1672-1675.

Körner, C. & Miglietta, F. (1994) Long term effects of naturally elevated CO_2 on mediterranean grassland and forest trees. *Oecologia*, **99**, 343-351.

Laisk, A. (1983) Calculation of photosynthetic parameters considering the statistical distribution of stomatal apertures. *Journal of Experimental Botany*, **149**, 1627-1635.

Longstreth, D.J. (1989) Photosynthesis and photorespiration in freshwater emergent and floating plants. *Aquatic Botany*, **34**, 287-299.

Miglietta, F. & Raschi, A. (1993) Studying the effect of elevated CO_2 in the open in a naturally enriched environment in central Italy. *Vegetatio* **104**, 290-301.

Miglietta, F., Raschi, A., Bettarini, I., R. Resti R. & Selvi, F. (1993) Natural CO_2 springs in Italy: A resource for examining the long-term response of vegetation to rising atmospheric CO_2 concentrations. *Plant, Cell and Environment*, **16**, 873-878.

Miglietta, F., Badiani, M., Bettarini, I., Van Gardingen, P., Selvi, F. & Raschi, A. (1994) Preliminary studies of the long-term response of Mediterranean vegetation around natural CO_2 vents. In *Anticipated effects of a changing global environment in Mediterranean-type ecosystems* (eds. J.M. Moreno & W.C. Oechel), pp. 102-120, Springer-Verlag, New York.

Morell, M.K., Paul, K., Kane, H.J. & Andrews, T.J. (1992) Rubisco: maladapted or misunderstood? *Australian Journal of Botany*, **40**, 431-441.

Morrison, J.I.L. (1987) Intercellular CO_2 concentration and stomatal response to CO_2. In *Stomatal function* (eds. E. Zeiger, G.D. Farquhar & I.R. Cowan), pp. 129-251, Stanford University Press, Stanford.

Mott K.A. (1990) Sensing of atmospheric CO_2 by plants. *Plant, Cell and Environment*, **13**, 731-737.

Nijs, I., Impens, I. & Van Hecke, P. (1992) Diurnal changes in the response of canopy photosynthetic rate to elevated CO_2 in a coupled temperature-light environment. *Photosynthesis Research*, **32**, 121-130.

Paoletti, E. & Gellini, R. (1993) Stomatal density variation in Beech and Holm Oak leaves collected over the past 200 years. *Acta Oecologia*, **14**, 173-178.

Peñuelas, J. & Matamala, R. (1990) Changes in N and S leaf content, stomatal density and specific leaf area of 14 plant species during the last three centuries of CO_2 increase. *Journal of Experimental Botany*, **41**, 1119-1124.

Pettersson, R. & McDonald, A.J.S. (1992) Effects of elevated carbon dioxide concentration on photosynthesis and growth of small birch plants (*Betula pendula* Roth.) at optimal nutrition. *Plant, Cell and Environment*, **15**, 911-919.

Sage, R.F., Sharkey, T.D. & Seemann, J.R. (1989) Acclimation of photosynthesis to elevated CO_2 in five C_3 species. *Plant Physiology*, **89**, 590-596.

Sage, R.F. (1994) Acclimation of photosynthesis to increasing atmospheric CO_2: The gas exchange perspective. *Photosynthesis Research*, **39**, 351-368.

Stitt, M. (1991) Rising CO_2 levels and their potential significance for carbon flow in photosynthetic cells. *Plant, Cell and Environment*, **14**, 741-762.

Vazzana, C., Puliga, S., Bochicchio, A. & Raschi, A. (1988) Studio della densità stomatica in un ecotipo di erba medica (*Medicago sativa* L.). *Rivista di Agronomia*, **22**, 127-132.

Von Caemmerer, S. & Farquhar, G.D. (1981) Some relationships between the biochemistry of photosynthesis and the gas exchange in leaves. *Planta*, **153**, 376-387.

Woodward, F.I. (1987) Stomatal numbers are sensitive to increases in CO_2 concentration from pre-industrial levels. *Nature*, **327**, 617-618.

Woodward, F.I., Thompson, G.B. & Mc Kee, I.F. (1991) The effects of elevated concentrations of carbon dioxide on individual plants, populations, communities and ecosystems. *Annals of Botany*, **67** (supplement 1), 23-38.

Carbon physiology of *Quercus pubescens* Wild. growing at the Bossoleto CO_2 spring in central Italy

J.D. JOHNSON[1], M. MICHELOZZI[2] AND R. TOGNETTI[2]

[1] *School of Forest Resources and Conservation, University of Florida, 326 Newins-Ziegler Hall, Gainesville, FL-32611, USA.*
[2] *Istituto Miglioramento Genetico delle Piante Forestali, Consiglio Nazionale delle Ricerche, via Atto Vannucci 13, 50134-Firenze, Italy.*

SUMMARY

The steady increase in the atmospheric concentration of carbon dioxide has important implications for the future growth and productivity of natural and managed ecosystems and is of particular interest to determine the carbon sinks useful for maintaining carboxylation efficiency in plant response to elevated CO_2 by eliminating the excess reduced carbon. The aim of this study was to evaluate the impact of exposure to naturally elevated CO_2 concentrations on the leaf carbon economy of a Mediterranean oak, *Quercus pubescens*. Measurements of net photosynthesis, leaf conductance to water vapour, transpiration, isoprene emission, and chlorophyll *a* fluorescence parameters were made on one-year-old seedlings (transplanted ten months prior to the experiment) and indigenous trees growing within the vicinity of a CO_2 spring and at an adjacent control site (4 km from the spring) on a clear day in August, 1993. Data on tree leaves were provided for comparison only. After measuring the leaves *in situ*, they were detached and allowed to dry for one hour, after which they were resampled for gas exchange and chlorophyll *a* fluorescence measurements. In addition, leaves from both seedlings and trees were sampled to enable specific leaf weight and tannin concentrations to be determined. Seedlings and the tree growing near the spring exhibited equal or slightly higher rates of photosynthesis, while leaf conductance was significantly lower, in comparison with plants growing at the control site. Instantaneous leaf water use efficiency was higher in plants growing near the spring than in control plants. In both seedlings and the tree growing at the spring site, rates of isoprene emission were about double those of control plants. In addition, both specific leaf weight and foliar tannin concentrations were significantly increased in the plants growing near the spring. Rapid desiccation of seedling leaves decreased all gas exchange and fluorescence parameters, and increased instantaneous water use efficiency; plants from the spring exhibited a 131 % increase in water use efficiency, while plants from the control site increased water use efficiency by only 20 %. This study indicates that many of the physiological responses to elevated CO_2 might be similar in both seedlings and older trees. Integrated measures of carbon physiology including specific leaf weight and tannin concentration were affected by CO_2 concentrations, and may have long-term significance for resistance to insect herbivory and litter decomposition rates. Higher rates of isoprene emission observed under elevated CO_2 may provide a mechanism for the plants to utilize excess reducing power that they are incapable of using for growth.

INTRODUCTION

Even though there is no general consensus on the degree and distribution of warming resulting from increased CO_2 in the atmosphere, or whether a global warming signal has already been detected, there is a growing interest in determining the effects of projected increased CO_2 on plant performance in conjunction with other stress factors such as increased water deficits and temperature (Chaves & Pereira, 1992).

Increasing concentrations of atmospheric CO_2 are generally expected to stimulate plant growth and productivity, but the extent of the effects may depend upon a number of factors such as species, nutrient status, temperature, water availability and pollution climate (Surano et al., 1986; Houpis et al., 1988). It is known that photosynthesis is enhanced by increases in ambient CO_2, at least in the short-term (Beadle et al., 1981; De Lucia, 1987; Tenhunen et al., 1984). At the cellular level, CO_2 acts as a substrate for ribulose bisphosphate carboxylase-oxygenase (rubisco) (Eamus & Jarvis, 1989) and therefore the level of CO_2 influences other enzymes that control plant metabolism and growth (Jarvis, 1989). It has been observed that stomatal conductance declines as CO_2 concentration increases and this results in a decreased rate of transpiration (Jarvis, 1989). At leaf level, enhanced CO_2 assimilation rate combined with decreased stomatal conductance leads to an increase in instantaneous water use efficiency.

After a long-term exposure to high CO_2, acclimated plants may show decreased rates of photosynthetic carbon assimilation (Eamus & Jarvis, 1989). Several hypotheses have been proposed to explain this phenomenon including starch accumulation and decrease in the amount and/or activity of rubisco (Mauney *et al.*, 1979; Peet, Huber & Patterson, 1985). However, the response to elevated CO_2 levels in acclimated trees can vary due to genetic variability and other environmental conditions.

To date, studies on plant-carbon dioxide interactions have focused on leaf gas exchange because of direct interactions between atmospheric CO_2 and its fixation in the leaf. However, changes in tissue chemistry, carbon allocation and nutrient use may influence gas exchange performance and growth. It is therefore of interest to evaluate the impact of CO_2 enhancement on secondary plant metabolism, particularly the allocation of carbon to foliar tannins for defence against herbivores (Swain, 1979), and to isoprene as a possible futile pathway for dealing with excess reduced carbon; in fact, carboxylation efficiency in plant response to elevated CO_2 seems to require the existence of active sinks (Jarvis, 1989) and isoprene may represent such a sink. In this study the primary and secondary carbon metabolism of *Quercus pubescens* growing at a naturally CO_2-enriched site in central Italy was characterized by measuring net photosynthesis, leaf conductance, transpiration, chlorophyll *a* fluorescence parameters, isoprene emissions, foliar tannin concentrations, and specific leaf weight. Measurements were taken on seedlings and indigenous trees grown under ambient and high CO_2 concentrations. Data on tree leaves were provided for comparison only.

Since, in the Mediterranean environment, one environmental change hypothesized to accompany the projected increase in atmospheric concentration of CO_2 is the increase in drought conditions, leaves of seedlings (which, even in normal conditions, may run into marked water stress) were also measured after being detached from the stem (Larcher, 1983; Reich 1984; Reich & Borchert, 1988; Schulte, Hinckley & Stettler, 1987). The purpose was to evaluate the resistance to rapid desiccation and compare the performances of leaves developed under ambient and elevated CO_2.

MATERIALS AND METHODS

The study was conducted at a natural CO_2 spring, Bossoleto, situated near Rapolano (Tuscany, central Italy). At this site, gas, low in H_2S (Miglietta et al., 1993), is emitted from a number of vents within a doline, resulting in a CO_2 concentration gradient ranging from 90 % to normal along the doline profile (Miglietta et al., 1993). The control site, 4 km from the spring, is characterized by similar soil and light exposure. Studies were conducted on one-year-old planted seedlings and indigenous trees of downy oak (*Quercus pubescens* Wild.) growing within the vicinity of the CO_2 spring and at the adjacent control site. The average CO_2 concentration during sampling was 518 ppm, though levels were considerably higher during the night and in the early morning. Average CO_2 concentrations at the control site were 379 ppm (significantly lower than at the spring). Seeds from a natural stand in central Tuscany were collected and grown in a mixture of local soil and sand at a tree nursery located at San Piero a Grado (Tuscany, central Italy). In October 1992, seedlings selected for uniformity were transplanted into the soil at both the spring and control sites, on a south-facing slope and near mature trees of the same species. Seedlings were irrigated regularly from the time of planting to supplement rainfall at the site, while trees received water only from rain. On August 5, 1993, nine seedlings plus one adjacent mature tree at each site were selected. A single leaf on each seedling and four leaves on each mature tree, at the same developmental stage, were selected for measurements (11:00-12:00 h, about 1500 μmol photons m^{-2} s^{-1}); the leaves of the seedlings were measured twice, first attached, and then one hour after detaching, in order to determine response to rapid water stress.

Net photosynthetic rate (A) was measured on one leaf per seedling and on four unshaded leaves per mature tree, using an infra-red gas analyser (CIRAS-1, PP Systems, Hitchin, UK) combined with an air supply unit (adjusted at a flow rate of 350 cm^3 min^{-1}) and a Parkinson broad-leaf leaf chamber (respectively, ASU-MF and PLC-B, Analytical Development Co. Ltd., Hoddesdon, UK). The photosynthetic rate was calculated from the difference in CO_2 content of in-going and out-going air and the flow rate of the air stream, according to the equations of Janac *et al.* (1971). Immediately after measuring the rates of CO_2 uptake, each leaf was also measured for leaf conductance (g_l) and transpiration (E) utilizing a null-balance porometer (LI-1600, Li-cor Inc., Lincoln, NE, USA). Instantaneous water use efficiency (WUE) was determined by combining photosynthesis and transpiration values.

Following the gas exchange measurements, chlorophyll *a* fluorescence induction kinetics were measured on the same leaves, using a portable fluorimeter (CF-1000, P.K. Morgan Instruments Inc., Andover, MA, USA). Leaves were dark acclimated for 15 min and were then illuminated with 800 μmol photons m^{-2} s^{-1} actinic light. Fast fluorescence kinetics were recorded over a 2 min period following illumination. The fluorescence parameters reported by the instrument are obtained with continuous illumination rather than the pulse-saturation technique. However, the time resolution of this fluorimeter during the onset of induction (20 μs point^{-1}) gives an accurate estimate of initial fluorescence (open or oxidized reaction centres) and, at higher PPFD, the actinic light is expected to fully reduce the primary quinone acceptor, providing an accurate estimate of the actual value of maximum fluorescence (closed or reduced reaction centres).

Isoprene emission rates were estimated on leaf disks (1.77 cm^2) taken from the same leaves used for gas exchange and chlorophyll *a* fluorescence measurements. The leaf disks were placed in separate 20 ml vials and incubated in full sun light (about 1500 μmol photons m^{-2} s^{-1}) for one hour in a shallow water-bath (Monson & Fall, 1989), and maintained at 28 °C. One ml of head-space gas was withdrawn with plastic syringes, transported to the laboratory (in the dark and on ice) and analysed by means of a gas chromatograph (AutoSystem, Perkin-Elmer Co., Norwalk, CO, USA) equipped with a flame ionization detector and a 15 m fused-silica (3 μm coating) column (Megabore DB-1, J&W Scientific, Folsom, CA, USA). Highly purified N_2 was used as carrier gas with a flow rate of 5 ml min^{-1} and gas was operated at a column temperature of 50 °C. Liquid isoprene with 99% purity (Aldrich Chemical Co., Milwaukee, WI, USA) was used for standardization and calibration purposes. The isoprene peak was identified by comparison of its retention time with that of authentic isoprene. Following isoprene analysis, leaf disks were dried in an oven at 70 °C for 48 h and specific leaf weight (SLW) was determined.

Tannin concentrations were measured on one leaf per seedling and on three leaves per tree. Leaves were sealed in plastic bags, placed on ice, transported to the laboratory and stored in a refrigerator at 4 °C. The next day, major veins were removed with a razor blade and discarded. The leaves were cut into small pieces; approximately 1 g of fresh leaf material was immediately ground in 5 ml of 70% ice-cold acetone using a tissue homogenizer, then the extracts were centrifuged at 5000 rpm for 15 min. The supernatants were stored at 4 °C in a refrigerator until analysis. Tannin concentrations were determined using a modified radial diffusion method of Hagerman (1987); a gel containing 1 % (w/v) agarose and 0.05 % (w/v) bovine serum albumin (BSA) in 50 mM acetic acid and 60 μM ascorbic acid adjusted to pH 5.0 (boiled and stirred, with the BSA added after cooling the agar to 45 °C) was formed by dispensing about 9.5 ml into standard plastic Petri dishes. Eight 5 mm wells per plate were punched in the gel and two 8 μl aliquots of each leaf extract were applied to duplicate wells. The Petri dishes were covered and sealed with Parafilm and incubated at 30 °C for 120 h. Two diameters at right angles of each diffusion ring were measured and the areas of the duplicate wells were averaged. Tannin concentration was determined using a standard curve created with purified tannic acid (Sigma Chemical Co., St. Louis, MO, USA).

The seedlings data were subjected to analysis of variance, using general linear model (GLM) procedures with Duncan's multiple range test used to determine probability at the 5 % level (SAS Institute Inc., Cary, NC, USA). Analysis of the tree samples provided an estimate of analytical error for trees data (Norby, Pastor & Melillo, 1986), but, because leaves were measured on one tree per site, there is no estimate of plant-to-plant variation and statistical analysis of the data is not possible.

RESULTS

The trees showed a lower overall photosynthetic rate than seedlings, possibly due to limited water for these plants or the often observed effect of tree age. The effect of CO_2 on photosynthesis in the plants was not evident although the seedlings exhibited higher but not significant photosynthetic rate at the spring site (Fig. 1A). The high CO_2 atmosphere affected leaf conductance of the experimental plants (Fig. 1A and B). Again, water stress and/or ageing in the trees were very evident in this parameter, as leaf conductance was between 5 and 10 times lower in the trees (Fig. 1A and B).

Transpiration rates were nearly identical in the seedlings, while the tree at the spring site showed a marked decrease (Fig. 1A and B). With respect to water use efficiency, transpiration appeared to dominate in the seedlings, where there was no CO_2 effect, while the spring tree exhibited an increase in water use efficiency (Fig. 1B).

Isoprene emission rates were significantly higher in the spring-grown seedlings, nearly double those of the control seedlings (Fig. 1A). The trees showed isoprene emission rates very similar to the seedlings and the spring tree exhibited an emission rate almost twice that of the control tree.

Chlorophyll a fluorescence parameters were quite similar between the seedlings and trees (Fig. 1C and D). The half-time rise for fluorescence induction (t1\2), a parameter which is a function of the rate of the photochemical reaction and the pool size of electron acceptors on the reducing side of photosystem II, was lower in both the seedlings and tree growing at the spring, the former being statistically different from the control site seedlings.

Relative fluorescence (F_v/F_m), which reflects the photochemical efficiency of open (oxidized) photosystem II reaction centres, was likewise depressed in plants at the spring site (Fig. 1C and D).

Specific leaf weight and tannin concentration showed no differences between seedlings and trees and were, therefore, combined for analysis (Fig. 2). In both instances, the plants growing in elevated CO_2 showed statistically higher values than plants at the control site. More than 1.25 mg per cm^2 of leaf area was detected in the leaves of the spring-grown plants compared to control plants. Likewise, foliar tannin concentrations were over 30 % higher in the spring plants than control plants.

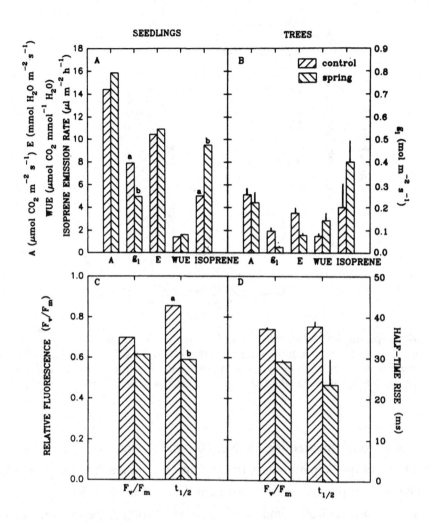

FIG. 1. Gas exchange and isoprene (panels A and B), and chlorophyll *a* fluorescence parameters (panels C and D) for *Quercus pubescens* seedlings (panels A and C) and trees (panels B and D) growing in either the CO_2 spring or control site. The parameters are: photosynthetic rate (A); transpiration (E); leaf conductance to water vapour (g_l); instantaneous water use efficiency (WUE); isoprene emission rate (ISOPRENE); relative fluorescence (F_v/F_m); half-time rise ($t_{1/2}$). Treatments are referred to by symbols in the legend. Significant mean separations by Duncan's multiple range test ($P < 0.05$) are indicated by different letters (seedlings data only). Bars represent ± SE of the mean ($n = 4$) (trees data only, see text for explanation).

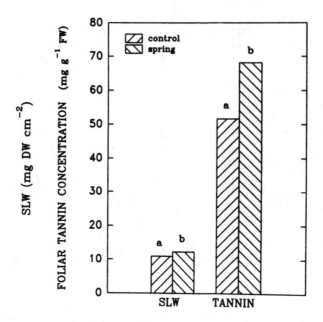

FIG. 2. Mean specific leaf weight (SLW) and foliar tannin concentrations of *Quercus pubescens* (averaged across seedlings and trees) growing in CO$_2$ spring and control site. Treatments are referred to by symbols in the legend. Significant mean separations by Duncan's multiple range test ($P < 0.05$) are indicated by different letters.

The response of the seedlings' leaves to rapid desiccation was dramatic and in a number of the parameters (transpiration, relative fluorescence and half-time rise) completely overwhelmed any CO$_2$ effect (Fig. 3).

In all instances, leaf removal decreased photosynthetic rate, leaf conductance, transpiration, relative fluorescence and half-time rise. Only in water use efficiency and isoprene emission did leaf detachment have little or no effect (Fig. 3).

Elevated CO$_2$ did interact with how rapid leaf desiccation affected photosynthesis, leaf conductance and water use efficiency (Fig. 4). Leaf desiccation had less of an impact on photosynthesis in the spring-grown seedlings (5.5 μmol CO$_2$ m^{-2} s^{-1}) than control seedlings (2.0 μmol CO$_2$m^{-2}s^{-1}), leading to a significant increase in the apparent water use efficiency of the spring-grown seedlings (131 % in spring *versus* 20 % in control seedlings). Reduction in leaf conductance was also less in the spring-grown seedlings primarily due to lower values of the attached leaves (Fig. 4B).

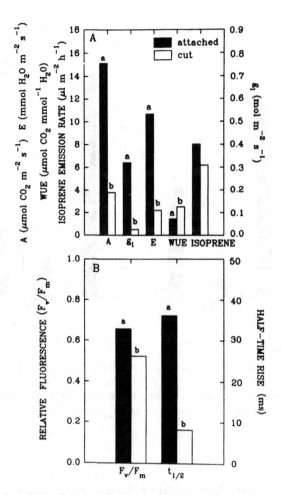

FIG. 3. The effect of rapid leaf desiccation (cut *versus* attached) on gas exchange (panel A) and chlorophyll *a* fluorescence (panel B) (see figure 1 for symbols). Parameters of *Quercus pubescens* seedlings averaged across CO_2 and control site. Treatments are referred to by symbols in the legend. Significant mean separations by Duncan's multiple range test ($P < 0.05$) are indicated by different letters.

DISCUSSION

Seedlings and trees of downy oak exhibited no appreciable variation in photosynthesis between sites though measurements taken on tannin concentrations, specific leaf weight and isoprene emissions suggest that the spring plants seem to have increased production of reduced carbon in response to elevated CO_2 by enhancing the production of these carbon-based secondary metabolites (see below).

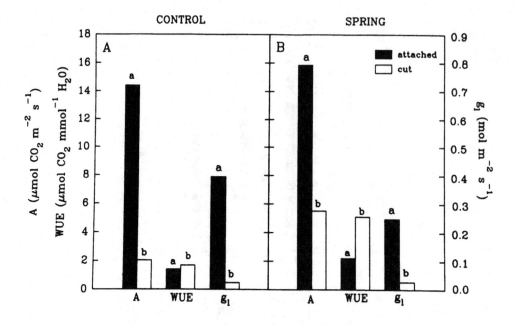

FIG. 4. The gas exchange (see Figure 1 for symbols) response of control (panel A) and CO$_2$ spring-grown (panel B) *Quercus pubescens* seedlings as affected by rapid leaf desiccation (cut *versus* attached). Treatments are referred to by symbols in the legend. Significant mean separations by Duncan's multiple range test ($P < 0.05$) are indicated by different letters.

Transpiration was significantly affected, resulting in a near doubling of water use efficiency in the tree growing in high CO$_2$. In contrast, water use efficiency remained relatively constant in the seedlings at the CO$_2$ spring.

The experimental plants grown in the spring had reduced half-time rise for fluorescence induction and relative fluorescence, indicating a decline in photochemical efficiency in response to high CO$_2$. A possible explanation could be that an adaptive alteration in their photosynthetic physiology occurred; plants growing in high CO$_2$ may not require elevated photochemical activity because of enhanced photosynthetic rates and reduced rates of photo-respiration (Eamus & Jarvis, 1989). The differences in gas exchange between seedlings and trees were principally due to the diverse watering regimes; the seedlings were irrigated regularly, while the trees suffered from water deficit due to seasonal drought, the spring tree being more water stressed than the control one (Tognetti et al., 1995).

The plants grown under high CO_2 exhibited lower stomatal conductance and transpiration rates. In general, stomatal conductance and transpiration rates decrease in response to increased CO_2 concentrations (Eamus & Jarvis, 1989). This effect is mediated by intercellular space CO_2, directly affecting the stomata (Mott, 1988). Intercellular space CO_2 depends upon the degree of coupling between assimilation rate and the ambient CO_2 concentration (Jarvis, 1989). However, the response of stomatal conductance to elevated CO_2 varies with species and environmental parameters including light, temperature and water supply (Jarvis, 1989; Eamus & Jarvis, 1989; Tolley & Strain, 1984a; Tschaplinski, Norby & Wullschleger, 1993).

A possible explanation for the observed higher isoprene emission from plants growing in elevated CO_2 is that isoprene production is a means by which excess reducing power is utilized. There is evidence that carboxylation efficiency in response to elevated CO_2 requires the existence of active carbon sinks (Koch et al., 1986) and isoprene biosynthesis may provide such a sink. The spring plants might have increased amounts of reduced carbon and release this surplus photosynthate through isoprene emission (Monson, Guenther & Fall, 1991) or by production of carbon-based secondary products (see below). The isoprene emission rate can account for between 1 and 8 % of the current rate of photosynthesis on a carbon basis (Tingey et al., 1979; Monson & Fall, 1989; Loreto & Sharkey, 1990). Sharkey, Loreto & Delwiche (1991) observed in Quercus rubra L. that over 15 % of current photosynthate is emitted as isoprene under high CO_2 and temperature (35 °C). Both of these conditions were present during our sampling at the spring site.

The increase in foliar tannin concentration could be the result of enhanced production of reduced carbon at high CO_2, where growth is limited by nitrogen availability. Körner & Miglietta (1994), performing chemical analysis of top-soils, found no differences in the nitrogen content between the control and the spring site. It is possible that available nitrogen does not match the higher atmospheric CO_2

concentration at the spring site. It is well known that carbon is primarily utilized in growth of the plant and excess carbon is used for the production of the carbon-based secondary metabolites (Tuomi *et al.*, 1988). This increase in secondary metabolites could affect the overall energy budget of the plant by increasing resistance to foliar herbivory, while decreasing growth (Jordan *et al.*, 1991). The observed increase in specific leaf weight is also consistent with utilization of excess carbon or a direct effect of CO$_2$ on leaf initiation (Eamus & Jarvis, 1989; Tolley & Strain, 1984b) and could have a long-term ecological impact by affecting litter decomposition rates.

Rapid leaf dehydration severely affected gas exchange parameters (photosynthesis, transpiration and leaf conductance) by about the same percentage in the seedlings of both sites. However, instantaneous water use efficiency increased by 20 % and 131 % in the control and spring seedlings, respectively. This is because leaf conductance was high in attached leaves of the control plants and leaf conductance in desiccated leaves from both sites dropped to 0.02 mol m^{-2} s^{-1}. In addition, cut leaves from spring plants had a higher photosynthetic rate than cut leaves from seedlings at the control site. Plants generally respond to elevated CO$_2$ by enhancing photosynthesis and decreasing stomatal conductance, resulting in increased instantaneous water use efficiency (Jarvis, 1989).

Chlorophyll *a* fluorescence parameters declined in response to rapid dehydration which would prevent photodamage when photosynthesis is inhibited by severe water stress conditions. A decrease in relative fluorescence (due mainly to reduced initial fluorescence, data not shown), i.e. an increase in non-photochemical deactivation of photosystem II, has been related to an increase in zeaxantin content promoting non-radiative dissipation (Demmig *et al.*, 1987). No significant differences were observed between seedlings from the two sites.

Isoprene emission was unaffected by rapid dehydration of leaves from seedlings at both sites, which is consistent with the suggestion that isoprene synthesis allows for the utilization of excess reducing power. Tingey, Evans & Gumpertez (1981) observed that live oak maintained a constant emission of isoprene even when gas exchange parameters were severely affected by water stress.

In conclusion, *Q. pubescens* responded positively to increased CO_2 concentration by fixing increased amounts of carbon resulting in higher foliar tannin concentrations, increased specific leaf weight and enhanced isoprene emission. In addition, the spring-grown plants exhibited low photochemical efficiency, suggesting a tendency towards acclimation of the photosynthetic apparatus to the elevated CO_2 atmosphere. The seedlings growing under high CO_2 responded better to rapid leaf dehydration than the control plants by increasing instantaneous water use efficiency more than twofold. Photo-inhibition occurred in seedlings at both sites in response to rapid leaf desiccation, suggesting that the photosynthetic apparatus of *Q. pubescens* leaves is sensitive to severe water stress. Isoprene emission was unaffected by rapid desiccation. Further research could reveal what stresses, if any, affect isoprene production in this important Mediterranean species. Overall, the seedlings responded to the short-term perturbations that we imposed similarly to the tree grown at the spring site. This is not surprising since the process of acclimation may require several hours to months in order to show a detectable response (Eamus & Jarvis, 1989). Due to the similar responses of the seedlings and the tree to elevated CO_2, further research is warranted to determine if *Q. pubescens* physiologically adapts to grow in the high CO_2 environment of the spring.

ACKNOWLEDGEMENTS

This research was supported by the National Research Council of Italy and a CNR-RAISA fellowship awarded to Jon D. Johnson. The authors wish to acknowledge Antonio Raschi (IATA-CNR, Italy) for his friendship and support during the study. The cost of the research has been partially covered by EU, Programme Environment, contract EV5V-CT93-0438.

REFERENCES

Beadle, C.L., Neilson, R.E., Jarvis, P.G. & Talbot, H. (1981) Photosynthesis as related to xylem water potential and CO_2 concentration in Sitka spruce. *Physiologia Plantarum*, **52**, 391-400.

Chaves, M.M. & Pereira, J.S. (1992) Water stress and climate change. *Journal of Experimental Botany*, **43**, 1131-1139.

De Lucia, E.H. (1987) The effect of freezing nights on photosynthesis, stomatal conductance and internal CO_2 concentration in seedlings of Engelmann spruce (*Picea engelmannii* Parry). *Plant, Cell and Environment*, **10**, 333-338.

Demmig, B., Winter, K., Krüger, A. & Czygan, F.C. (1987) Photoinhibition and zeaxanthin formation in intact leaves. A possible role of the xanthophyll cycle in the dissipation of excess light energy. *Plant Physiology*, **84**, 218-224.

Eamus, D. & Jarvis, P.G. (1989) The direct effects of increase in the global atmospheric CO_2 concentration on natural and commercial temperate trees and forests. *Advances in Ecological Research*, **19**, 1-55.

Hagerman, A.E. (1987) Radial diffusion method for determining tannin in plant extracts. *Journal of Chemical Ecology*, **13**, 437-449.

Houpis, J.L.J., Surano, K.A., Cowles, S. & Shinn, J.H. (1988) Chlorophyll and carotenoid concentrations in two varieties of *Pinus ponderosa* seedlings subjected to long-term elevated carbon dioxide. *Tree Physiology*, **4**, 187-193.

Janac, J., Catsky, J., Brown, K.J. & Jarvis, P.G. (1971) Gas handling system. In *Plant photosynthetic production: manual and methods* (eds. Z. Sestak, J. Catsky & P.G. Jarvis), pp. 132-148. Dr. Junk, The Hague.

Jarvis, P.G. (1989) Atmospheric carbon dioxide and forests. *Philosophical Transactions of the Royal Society of London*, **B 324**, 369-392.

Jordan, D.N., Green, T.H., Chappelka, A.H., Lockaby, B.G., Meldahl, R.S. & Gjerstad, D.H. (1991) Response of total tannins and phenolics in loblolly pine foliage exposed to ozone and acid rain. *Journal of Chemical Ecology*, **17**, 505-513.

Koch, K.E., Jones, P.H., Avigne, W.T. & Allen, L.H. (1986) Growth, dry matter partitioning, and diurnal activities of RuBP carboxylase in *Citrus* seedlings maintained at two levels of CO_2. *Physiologia Plantarum*, **67**, 477-484.

Körner, Ch. & Miglietta, F. (1994) Long term effects of naturally elevated CO_2 on mediterranean grassland and forest trees. *Oecologia*, **99**, 343-351.

Larcher, W. (1983) *Physiological plant ecology*. Springer -Verlag, Berlin.

Loreto, F. & Sharkey, T.D. (1990) A gas-exchange study of photosynthesis and isoprene emission in *Quercus rubra* L.. *Planta*, **182**, 523-531.

Mauney, J.R., Guinn, G., Fry, K.E. & Hesketh, J.D. (1979) Correlation of photosynthetic carbon dioxide uptake and carbohydrate accumulation in cotton soybean, sunflower and sorghum. *Photosynthetica*, **13**, 260-266.

Miglietta, F., Raschi, A., Bettarini, I., Resti, R. & Selvi, F. (1993) Natural CO_2 springs in Italy: a resource for examining long-term response of vegetation to rising atmospheric CO_2 concentrations. *Plant, Cell and Environment*, **16**, 873-878.

Monson, R.K. & Fall, R. (1989) Isoprene emission from Aspen leaves. The influence of environment and relation to photosynthesis and photorespiration. *Plant Physiology*, **90**, 267-274.

Monson, R.K., Guenther, A.B. & Fall, R. (1991) Physiological reality in relation to ecosystem and global level estimates of isoprene emission. In *Trace gas emission by plants* (eds. T.D. Sharkey, E.A. Holland & H.A. Mooney), pp.185-207. Academic Press, San Diego.

Mott, K.A. (1988) Do stomata respond to CO_2 concentration other than intercellular? *Plant Physiology*, **86**, 200-203.

Norby, R.J., Pastor, J., Melillo, J.M. (1986) Carbon-nitrogen interactions in CO_2-enriched white oak: physiological and long-term perspectives. *Tree Physiology* **2**, 233-241.

Peet, M.M., Huber, S.C. & Patterson, I.T. (1985) Acclimation to high CO_2 in *Monoecious cucumber*. II. Alterations in gas exchange rates, enzyme activities and starch and nutrient concentrations. *Plant Physiology*, **80**, 63-67.

Reich, P.B. (1984) Loss of stomatal function in aging hybrid poplar leaves. *Annals of Botany*, **53**, 691-698.

Reich, P.B. & Borchert, R. (1988) Changes with leaf age in stomatal function of several tropical species. *Biotropica*, **20**, 60-69.

Schulte, P.J., Hinckley, T.M. & Stettler, R.F. (1987) Stomatal response of Populus to leaf water potential. *Canadian Journal of Botany*, **65**, 255-260.

Sharkey, T.D., Loreto, F. & Delwiche, C.F. (1991) High carbon dioxide and sun/shade effects on isoprene emission from oak and aspen tree leaves. *Plant, Cell and Environment*, **14**, 333-338.

Surano, K.A., Daley, P.F., Houpis, J.L.J., Shinn, J.H., Helms, J.A., Palassou, R.J. & Costella, M.P. (1986) Growth and physiological responses of *Pinus ponderosa* Doug. ex P. Laws. to long-term elevated CO_2 concentrations. *Tree Physiology*, **2**, 243-259.

Swain, T. (1979) Tannins and lignins, In *Herbivores: their interaction with secondary plant metabolites* (eds. G.A. Rosenthal & D.H. Janzen), pp. 657-682, Academic Press, New York.

Tenhunen, J.D., Lange, O.L., Gebel, J., Beyschlag, W. & Weber, J.A. (1984) Changes in photosynthetic capacity, carboxylation efficiency, and CO_2 compensation point associated with midday stomatal closure and midday depression of net CO_2 exchange of leaves of *Quercus suber*. *Planta*, **162**, 193-203.

Tingey, D.T., Manning, M., Grothaus, L.C. & Burns, W.F. (1979) The influence of light and temperature in isoprene emission rates from live oaks. *Physiologia Plantarum*, **47**, 112-118.

Tingey, D.T., Evans, R. & Gumpertez, M. (1981) Effects of environmental conditions on isoprene emission from live oak. *Planta*, **152**, 565-570.

Tognetti, R., Giovannelli, A., Longobucco, A., Miglietta, F. & Raschi, A. (1995) Water relations of oak species growing in the natural CO_2 spring of Rapolano (Central Italy). *Annales des Sciences Forestières*, **53**, 475-485.

Tolley, L.C. & Strain, B.R. (1984a) Effects of CO_2 enrichment on growth of *Liquidambar styraciflua* and *Pinus taeda* seedlings under different irradiance levels. *Canadian Journal of Forest Research*, **14**, 343-350.

Tolley, L.C. & Strain, B.R. (1984b) Effects of CO_2 enrichment and water stress on growth of *Liquidambar styraciflua* and *Pinus taeda* seedlings. *Canadian Journal of Botany*, **62**, 2135-2139.

Tschaplinski, T.J., Norby, R.J. & Wullschleger, S.D. (1993) Responses of loblolly pine seedlings to elevated CO_2 and fluctuating water supply. *Tree Physiology*, **13**, 283-296.

Tuomi, J., Niemelä, P., Chapin III, F.S., Bryant, J.P. & Sirén, S. (1988) Defensive responses of trees in relation to their carbon/nutrient balance. In *Mechanisms of woody plant defenses against insects: search for pattern* (eds. W.J. Mattson, J. Levieux & C. Bernard-Dagan), pp. 57-72. Springer Verlag, New York.

Preliminary results on dissolved inorganic ^{13}C and ^{14}C content of a CO_2-rich mineral spring of Catalonia (NE Spain) and of plants growing in its surroundings

J. PIÑOL AND J. TERRADAS

CREAF, Universitat Autònoma de Barcelona, 08193 Bellaterra, Spain.

SUMMARY

A preliminary survey for CO_2-rich mineral springs around the area of Girona (NE Spain) was conducted having in mind their possible use for biological research. At present, the survey has produced only two mineral springs that create a CO_2 enriched atmosphere and that have natural vegetation around them. One of them (Sant Hilari) was further studied in order to discover (1) if the CO_2 degassed from a spring has a different ^{13}C and ^{14}C signature from that of atmospheric CO_2; (2) if plants growing in the surroundings of the spring have a distinctive C isotopic composition from those growing under a normal CO_2 atmosphere; and (3) if it would be possible to back-estimate the mean CO_2 concentration that plants have experienced during their lifetime, from considering plant C isotopic content as a mixture of two end-members, air CO_2 and spring-derived CO_2.

Results showed that: - Dissolved inorganic carbon of springwater was slightly enriched in ^{13}C (-1.6‰ PDB) and very depleted in ^{14}C (3.15% of modern carbon) compared to atmospheric CO_2 (-8.0‰ PDB ^{13}C and 100% ^{14}C).

- The isotopic signal of ^{14}C of CO_2 spring was detected in plant materials but not that of ^{13}C. This result corresponds to the much higher difference between the two end-members (spring and atmospheric CO_2) compositions for ^{14}C (near 100%) than for ^{13}C (only 6.4‰).

- Consequently, it is possible to estimate from ^{14}C, but not from ^{13}C, analysis that the mean CO_2 concentration that *Angelica sylvestris* (Apiaceae) plants growing around the spring have experienced during their lifetime was 405 ppmv.

INTRODUCTION

Natural CO$_2$ springs create in their surroundings a CO$_2$-enriched atmosphere. CO$_2$ can reach the surface as a gas emanation (dry springs) or dissolved in water (CO$_2$-rich mineral springs). In principle, CO$_2$ springs might be used for testing predictions about changes in plant communities in response to elevated CO$_2$ in natural conditions maintained over long periods of time. The effect of these springs on plants has been studied in very few places of the world, such as Italy (Miglietta & Raschi, 1993), Iceland (Cook *et al.*, this volume), and California (Koch), but they likely exist in many other places.

Nevertheless, there are some problems associated with the physical characteristics and location of these springs that must be taken into account before considering CO$_2$ springs as a useful tool for biological research, especially if the results are to be compared with controlled environment experiments. One of the problems is that the physical location of the springs and the variable weather conditions, especially wind speed and direction, make the CO$_2$ concentration highly fluctuating. As a consequence, it is difficult to assign a specific CO$_2$ concentration to a given location.

Obviously, CO$_2$ concentration can be continuously monitored, but this is expensive and time-consuming. An alternative method is to know the CO$_2$ concentration that a plant has experienced during its lifetime based on the use of stable carbon isotopes.

If the CO$_2$ degassed from springwater has a different C isotopic signature than that of the atmospheric CO$_2$, this characteristic may be used to trace the fate of CO$_2$ from springwater to plant material, and to estimate the mean CO$_2$ concentration that plants have experienced during their lifetime.

The detailed objectives of this study were: (1) To find in NE Spain CO$_2$-rich mineral springs that could be used for biological research. (2) To discover if the CO$_2$ degassed from a spring has a different ^{13}C and ^{14}C signature from that of atmospheric CO$_2$. (3) If affirmative, to discover if plants naturally growing in the surroundings of the spring have a distinctive C isotopic composition from those growing under the normal CO$_2$ atmosphere. This would mean that these plants have been growing in an elevated CO$_2$ atmosphere. (4) Finally, if part of the CO$_2$ degassed from the spring has been incorporated into plant materials, and this resulted in a distinctive isotopic composition, from it could be possible to back-estimate the mean CO_2 concentration that plants have experienced during their lifetime.

SPRINGS SURVEY AND SITE DESCRIPTION

We have surveyed the area of Girona (NE Spain) for mineral springs that create a CO_2 enriched atmosphere and that are surrounded by natural vegetation. At present, we have only found two of these springs. This study deals with one of them; the second one has been located only recently. The studied spring is located near the village of Sant Hilari de Sacalm at an altitude of 690 m a.s.l. Its geographical coordinates are 41°54' 26" N and 2°30' 46" E. It has been modified by man, but water leaving the outlet of the spring circulates through natural vegetation near a five meter-long stream. The spring discharge is very low (around 1 l min^{-1}). The more abundant plant species growing near the spring are: *Hedera helix*, *Ranunculus acris*, *Urtica dioica*, *Viola* sp., *Equisetum telmateia*, *Carex remota*, *Geranium robertianum*, and *Angelica sylvestris* (Phanerophytes), and *Plagiomnium undulatum*, *Conocephalum conicum*, and *Pellia endiviifolia* (Bryophytes).

MATERIALS AND METHODS

CO_2 concentration in the air was measured using short-term Dräger tubes. Springwater was chemically analyzed for pH (electrometrically), alkalinity (Gran titration), SO_4^{2-}, Cl^- and NO_3^- (ion chromatography), Na^+, K^+, Ca^{2+}, and Mg^{2+} (atomic spectrometry), and Fe^{2+} (coupled induction plasma). Equilibrating water pCO_2 and dissolved inorganic carbon (DIC) were calculated from alkalinity and pH measurements using accepted equations and equilibrium constants of the carbonic system (Stumm & Morgan, 1981).

The entire aerial part of *Angelica sylvestris* plants were harvested at distances of 0-1 m and 7 m from the spring outlet. Leaves of *Hedera helix* were also collected at three distances ranging from 0.1 to 13 m from the spring outlet. Carbon isotopes in water DIC and in plant materials were analyzed in the following way: DIC was precipitated with $BaCl_2$ after raising the water pH with NaOH. Plants were oven dried (70°C) and ground. ^{13}C of the precipitated $BaCO_3$ and of plant materials was analyzed on a ratio mass spectrometer (VG-Isogas, Cheshire, UK), and ^{14}C by the liquid scintillation counting technique in a liquid scintillation counter (LKB Wallac 1220 Quantulus, Turku, Finland). ^{13}C isotope ratios are reported as $\delta^{13}C$ values referred to the PDB standard. ^{14}C concentrations are reported as a percentage of modern carbon (%M).

The spring was visited three times during 1993. Springwater for chemical and isotopic analysis was collected on 18 June. Atmospheric CO_2 concentrations were measured on 13 February and 29 September. Plants for C isotopic analysis were sampled on 13 February and 18 June.

RESULTS

Water chemistry and air CO_2 concentration

The chemical composition of Sant Hilari springwater (Table 1), like other similar springs of the area, was highly mineralized and very rich in DIC. The water was oversaturated with CO_2, and when it comes in contact with the open atmosphere a net flux of CO_2 is established from the springwater to the air. Consequently, the air next to the spring outlet is enriched in CO_2 compared to the open atmosphere.

Nevertheless, the pattern of CO_2 concentrations in the proximity of the spring was highly variable both in space and in time. For instance, on 13 February 1993 CO_2 concentration just at the spring outlet was over 6000 ppmv; 75 cm from the outlet, it was still 750 ppmv. On 29 September 1993, a windy day, values close to the outlet fluctuated between 600 and 1800 ppmv, whereas even at a short distance (25 cm) from the water surface CO_2 concentration was already at the atmospheric level.

TABLE 1. Chemical composition of Sant Hilari springwater on 18 June 1993. Concentrations are expressed in mEq L^{-1} (except DIC that is in mmol L^{-1}), electrical conductivity (E.C.) in mS cm^{-1} at 25°C, and pCO_2 in atm.

Na^+	3.50	HCO_3^-	21.9
K^+	0.03	Cl^-	0.30
Ca^{2+}	16.3	SO_4^{2-}	0.16
Mg^{2+}	2.33	NO_3^-	ind.
Fe^{2+}	1.80		
Sum of cations	24.0	Sum of anions	22.4
pH	6.46	E.C	1573
DIC	39.3	pCO_2	0.38

C isotopic content of DIC and of atmospheric CO_2

^{13}C and ^{14}C content of DIC springwater were -1.6‰ and 3.15±0.11%M, respectively. These values are slightly different from those of atmospheric CO_2 for ^{13}C (-8‰; Farquhar, Ehleringer & Hubick, 1989a) and very different for ^{14}C (100%M).

Consequently, the CO_2 enriched air surrounding the spring outlet must have a different C isotopic composition than the atmospheric CO_2, and this difference will depend on the relative proportion of CO_2 degassed from the spring and from the open atmosphere (Table 2).

C isotopic content of plants

(1) Leaves and stems of A. sylvestris and H. helix were depleted in ^{13}C (Table 3) compared to the two possible sources (air and spring) of CO_2. The magnitude of the depletion was within the known range for C3 plants (Farquhar et al., 1989a and b).

(2) Differences in 13C composition between the plants growing near the spring outlet and further away were very small (Table 3).

TABLE 2. Theoretical C isotopic composition of CO_2 of the air affected by the degassing of the spring. C isotopic composition of the air mixtures enriched in CO_2 was calculated from the C isotopic contents of the two end-members, air and spring-derived CO_2. It was assumed that atmospheric CO_2 concentration is 350 ppmv and that C isotopic fractionation during the degassing process is negligible (Farquhar, Ehleringer & Hubick, 1989a).

End-members			$\delta^{13}C$	^{14}C
			(‰ PDB)	(% M)
Atmospheric CO_2			-8.0	100
Spring CO_2			-1.6	3.15
Mixtures				
Atmos.	Spring	Mixture	$\delta^{13}C$	^{14}C
350	0	350 ppmv	-8.0	100
350	55	405 ppmv	-7.1	87
350	150	500 ppmv	-6.1	71
350	650	1000 ppmv	-3.8	37
350	4650	5000 ppmv	-2.0	10

(3) Nevertheless, the ^{14}C of A. sylvestris leaves growing near (0-1 m) the spring outlet retained partly the ^{14}C signature of the spring-derived CO_2 (82.4%M). Considering that the isotopic discrimination against ^{14}C in photosynthesis is -4.4% (assumed to be twice that of ^{13}C, -2.2%, Farquhar et al., 1989a and b), then the A. sylvestris leaves growing near the spring outlet have been exposed to CO_2 with a mean ^{14}C content of around 87% (calculated as ^{14}C of A. sylvestris leaves, 82.4%, corrected for discrimination against ^{14}C, 4.4%). This implies that the mean concentration of atmospheric CO_2 experienced by the plants was approximately 405 ppmv (Table 2). The ^{14}C of A. sylvestris leaves growing 7 m away was totally modern, showing that the enrichment effect of the spring CO_2 is negligible at such a distance.

TABLE 3. ^{13}C isotopic composition of plants growing near the carbon dioxide spring of Sant Hilari. Distance is referred to the spring outlet.

Species	Distance	δ^{13}C PDB
18 June 1993		
Angelica sylvestris stems	0-1 m	-29.9‰
Angelica sylvestris leaves	0-1 m	-30.1‰
Angelica sylvestris stems	7 m	-30.6‰
Angelica sylvestris leaves	7 m	-30.0‰
Hedera helix leaves	0.4 m	-30.2‰
Hedera helix leaves	2.5 m	-28.1‰
Hedera helix leaves	13 m	-30.6‰
13 February 1993		
Hedera helix leaves	0.1 m	-31.2‰
Hedera helix leaves	0.5 m	-31.3‰
Hedera helix leaves	10 m	-30.7‰

DISCUSSION

The CO_2-rich mineral spring of Sant Hilari, and probably some other springs of the same geographical area, extend the number of sites in the world where there are CO_2 springs that can be used in biological research. Nevertheless, the Sant Hilari spring is much smaller and produces much less CO_2 than other well-known CO_2 springs, like those of Tuscany in central Italy (van Gardingen *et al.*, this volume; Paoletti *et al.*, this volume; Miglietta & Raschi, 1993). But as the present research shows the usefulness of CO_2 springs to understand long-term effects of enriched CO_2 atmospheres on plants, then it would be extremely useful to have as many and as diverse study sites as possible. In that case it would be worthwhile to study even the smaller springs.

The DIC of the studied spring, and presumably also the degassed CO_2, has a different C isotopic signature than the atmospheric CO_2. The C isotopic difference between atmospheric and spring derived CO_2 is much more noticeable for ^{14}C than for ^{13}C. In the Sant Hilari spring it was shown that ^{14}C, but not ^{13}C, content of plant material can be used to estimate the mean CO_2 concentration that plants have experienced during their lifetime. The fact that the isotopic signal of ^{14}C of the CO_2 spring was detected in plant materials but not that of ^{13}C, corresponds to the much higher difference between the two end-members' (springwater DIC and atmospheric CO_2) compositions for ^{14}C (near 100%) than for ^{13}C (only 6.4‰). This result contrasts with that shown by Koch where leaf ^{13}C content varies with proximity to the spring. The difference between Koch's results and ours arises probably from the lower CO_2 concentrations in the air of the Sant Hilari spring than in the Clear Lake spring studied by Koch. As the difference between air and spring-derived CO_2 is usually only of a few per mil units, the isotopic signature of the spring CO_2 can only be detected in plants when the mean air CO_2 concentration is relatively high.

From a methodological point of view it would be much better to use ^{13}C than ^{14}C as the isotopic tracer of spring CO_2: ^{13}C analysis is one order of magnitude cheaper and requires two orders of magnitude less plant material than ^{14}C analysis (by the liquid scintillation method). The amount of plant material needed for ^{14}C analysis is limiting in particular when the area of increased CO_2 around the spring is small. In Sant Hilari spring there was only one species (*Angelica sylvestris*) present in a quantity large enough to obtain the several grams (at least 3 g of C) of leaf material required for ^{14}C analysis. In addition, to get it we had to clear out an area of 1 m^2 between the spring outlet and 1 m away, losing the possibility of detecting possible CO_2 concentration gradients in the proximity of the spring.

A second CO_2-rich mineral spring (Sant Gregori) recently located seems to be more adequate for biological research. Preliminary results were promising: spring discharge and DIC are higher than in Sant Hilari. The spring is located in a concavity that decreases ventilation, and there are several plant species (*Rubus ulmifolius, Hedera helix*) present in large amounts.

ACKNOWLEDGEMENTS

David Bonilla, Montserrat Brugués, Josep Girbal, Ramon Vaquer, Josep Trilla, Joaquim Merín, Isidre Casals, Colin Neal, Esther Granados, and Carme Melcion are gratefully acknowledged for their technical support in the field or in the laboratory. Ferran Rodà critically revised the manuscript. Marianne Mousseau and another reviewer also contributed to improve the paper.

REFERENCES

Farquhar, G.D., Ehleringer, J.R. & Hubick, K.T. (1989a) Carbon isotope discrimination and photosynthesis. *Annu. Rev. Plant Physiol. Plant Mol. Biol.*, **40,** 503-537.

Farquhar, G.D., Hubick, K.T., Condon, A.G., & Richard, R.A. (1989b) Carbon isotope fractionation and plant water-use efficiency. In *Stable Isotopes in Ecological Research* (eds. Rundel, P.W., Ehleringer, J.R., & Nagy, K.A.). Springer-Verlag. New York.

Miglietta, F. & Raschi, A. (1993) Studying the effect of elevated CO_2 in the open in a naturally enriched environment in Central Italy. *Vegetatio,* **104-105,** 391-400.

Stumm, W. & Morgan, J.J. (1981) *Aquatic Chemistry.* John Wiley & Sons. New York.

The impact of elevated CO_2 on the growth of *Agrostis canina* and *Plantago major* adapted to contrasting CO_2 concentrations

M.C. FORDHAM[1], J.D. BARNES[1], I. BETTARINI[2],
H.G. GRIFFITHS[1], F. MIGLIETTA[2] AND A. RASCHI[2]

[1]*Department of Agricultural and Environmental Science,*
Ridley Building, University of Newcastle upon Tyne,
Newcastle-upon-Tyne, NE1 7RU, U.K.
[2]*IATA-CNR (Institute of Agrometeorology and Environmental Analysis,*
Consiglio Nazionale delle Ricerche),
P. le delle Cascine, 18 50144 Firenze, Italy.

SUMMARY

Plant species native to CO_2 springs have evolved along gradients of naturally elevated atmospheric CO_2 concentrations that differ as much as predicted for anthropogenic increases in CO_2 over the next 100-200 years. This study characterized differences in gro wth and dry matter partitioning under controlled conditions in two species, *Agrostis canina* L. spp. *montelucci* (Selvi, 1994) and *Plantago major* L., adapted to contrasting atmospheric CO_2 concentrations. When plants were grown at twice the present CO_2 level there was a pronounced increase in biomass production, which was largely due to an initial stimulation in relative growth rate. In *Agrostis canina*, CO_2 enrichment caused an increase in the relative partitioning of dry matter to the root, but effects were less pronounced in *Plantago major*. Specific leaf area was reduced by elevated CO_2 in both species. Plants adapted to growth at high CO_2 produced greater biomass and exhibited higher initial RGRs than plants adapted to lower CO_2 concentrations, and in the case of *Agrostis canina* these differences were maintained at ambient CO_2. Differences in original seed weight between populations were small (<10% for *Agrostis* and < 20% for *Plantago*). A positive correlation was, however, found between seed weight and the long-term atmospheric CO_2 concentration at the site of collection and larger seeds were associated with higher initial plant RGRs under CO_2 enriched conditions. Yet, differences in seed weight could explain no more than 33% and 43% of the variation in initial RGR between populations of *Agrostis* and *Plantago*, respectively. These results suggest that, at least initially, plants originating from high-CO_2 have an intrinsically greater growth potential than plants originating from ambient CO_2, which may be partly explained by the production of larger seeds. However, other traits as yet unidentified appear to be largely responsible for the intrinsic differences in growth potential observed in plants originating from contrasting long-term atmospheric CO_2 concentrations.

INTRODUCTION

Atmospheric concentrations of carbon dioxide (CO_2) are rising at an unprecedented rate (Conway *et al.,* 1988) and this has prompted considerable interest in the potential impacts of elevated CO_2 on natural and managed ecosystems. Research has shown that a number of factors including increased rates of CO_2 assimilation, higher water-use efficiency, decreased respiration, and the relief of nutrient stress *via* enhanced nutrient use efficiency and nutrient uptake will contribute to increased plant growth and productivity in the short-term (Eamus & Jarvis, 1989; Allen, 1990; Kimball *et al.*, 1990). There remains, however, considerable uncertainty over the long-term responses of vegetation to increased CO_2 under field conditions.

Many controlled environment, open-top chamber and glasshouse studies have revealed that CO_2-induced stimulations in growth and photosynthesis may not be sustained under conditions of prolonged CO_2 enrichment, because plants "acclimate" to elevated CO_2 (see Stitt, 1991; Arp, 1991). Yet there is growing evidence that such "acclimation" responses are not inevitable (Petterson & McDonald, 1994) and many studies in which well-fertilized plants have been grown in the field or in containers large enough to prevent root restriction have shown little or no decrease in photosynthetic capacity in response to long-term CO_2 enrichment (Arp, 1991; Barnes, Ollerenshaw & Whitfield, 1995). As a consequence, it is now considered that in most agricultural situations, where the soil is well tilled and nutrient supply plentiful, many crop plants will continue to respond to increased CO_2 over their entire life-span (Kimball *et al.*, 1990). Much less information is available on which to base predictions of the long-term response of natural plant communities to increased atmospheric CO_2 concentrations, but it is believed that soil compaction combined with limited supplies of other resources (most importantly nutrients, water and light) may reduce the potential growth enhancement resulting from increased CO_2 (Grulke *et al.*, 1990; Drake & Leadley, 1991; Petterson & McDonald, 1994).

Native plant material has a considerably greater ecological and physiological range than cultivated species so the long-term responses of natural plant communities to elevated CO_2 cannot necessarily be predicted from experiments on highly selected crop plants grown under nutrient-rich conditions (Maxon-Smith, 1977). There are many reports showing that considerable variation exists between and within species in response to CO_2 enrichment (Wulff & Alexander, 1985; Strain & Cure, 1986). Strain (1991) suggests this diversity may represent residual genetic potential for increased growth and competition under CO_2 enriched conditions, since many plant species evolved under conditions where the atmospheric CO_2 concentration was considerably higher than it is at present. Given the potential existence of a pool of heritable variation in CO_2 responsiveness within plant species, it has been suggested that anthropogenic increases in the atmospheric CO_2 concentration may act differentially on plant populations with the result that genotypes better able to make use of the available resources will prosper and increase in frequency to the detriment of plants unable to do so and lead to eventual genetic shifts within plant populations (Strain, 1987; Friend & Woodward, 1990). It has not, however, proved possible to test this hypothesis because of the fundamental discrepancies between the time scales of experiments and the time scales relevant in an ecological context. To date, the longest controlled exposures of natural vegetation to elevated CO_2 have lasted for no more than four years (Grulke *et al.*, 1990; Drake & Leadley, 1991), yet periods of 10 years or more will be necessary to establish how the genetic structure of even short-lived species changes in response to increased CO_2 (Miglietta *et al.*, 1993a). In the meantime it has been suggested that one way of providing a more realistic picture of long-term vegetation responses to CO_2 is to identify specific characteristics or physiological traits existing in plants originating from those few special locations where atmospheric CO_2 concentrations have been naturally elevated over evolutionary time scales. Such sites include geothermal CO_2 springs (Miglietta *et al.*, 1993a, 1993b), CO_2-rich microsites on respiring substrates (Sonesson, Gehrke & Tjus, 1992) or by comparing plants from contrasting altitudes (Friend & Woodward, 1990; Körner & Diemer, 1994).

This paper reports the results of a study examining the effects of pre-selection in a high-CO_2 environment on growth and dry matter partitioning in two species, *Agrostis canina* L. spp. *montelucci* (Selvi, 1994) and *Plantago major* L., maintained under controlled conditions at ambient and elevated CO_2. Seed of both species was collected from plants growing at contrasting atmospheric CO_2 concentrations within the vicinity of two natural geothermal springs in central Italy. At these sites, as at other CO_2 springs in the region, the native vegetation has been exposed for many generations to naturally elevated concentrations of CO_2 *via* geothermal vents that emit almost pure CO_2 (93-97%) to the atmosphere. One major advantage of these sites is the existence of long-term-average atmospheric CO_2 concentration gradients with increasing distance from the emission source, enabling direct comparisons to be made between plants adapted to contrasting atmospheric CO_2 concentrations.

MATERIALS AND METHODS

Plant material and collection site details
Agrostis canina

In July 1993 seed of *A. canina* was collected from plants growing along a natural gradient of CO_2 concentrations at a CO_2 spring (Solfatara) located near Viterbo in central Italy (Miglietta & Raschi, 1993; Miglietta *et al.*, 1993a; Selvi, 1996). At this site a number of vents are dispersed over an area of about 2 ha and the area is dissected by several small streams. The gas emitted by the vents is relatively pure CO_2 (93%) but also contains trace amounts of oxygen, methane, oxides of nitrogen and hydrogen sulphide (Duchi, Minissale & Romani, 1985). As a result of the gas emissions there is significant enhancement of atmospheric CO_2 concentrations over a large area and a gradient of decreasing CO_2 concentrations with increasing distance from the major vents - although short-term fluctuations in CO_2 concentrations due to atmospheric turbulence may affect the extent of this gradient, particularly at night (Miglietta & Raschi, 1993; Miglietta et al., 1993a). Leaf nutrient analyses, and soil physical and chemical characteristics at the site show no correlations with atmospheric CO_2 concentration at the vent site (Fordham and Barnes, unpublished results).

Because of the technical difficulties encountered in the direct determination of long-term average CO_2 concentrations by infra-red gas analysis at this, and other CO_2 springs, average long-term integrated growth CO_2 concentrations along the transect were determined from the stable carbon isotope composition of C_4 plant material - under the assumption that discrimination against $^{13}CO_2$ remains constant in C_4 plants under conditions of long-term CO_2 enrichment. Seed of pearl millet (Pennisetum americanum), maize (Zea mays) and sorghum (Sorghum bicolor) was sown in pots and placed along a 160 m transect at Solfatara in late spring 1992 and 1993, and watered daily. In 1992 leaves of these plants were harvested three months after sowing, and in 1993 at three time intervals (1, 2 and 3 months after sowing). Leaf material was immediately oven-dried, finely ground, and the relative abundance of ^{13}C and ^{12}C in combusted samples subsequently determined using a dual inlet mass spectrometer (I. Bettarini, personal communication). The isotopic ratio of the CO_2 emitted from the vents at Solfatara has a characteristic signature (-0.2‰) which is considerably lower than that of "normal" air (-8‰) and was reflected in significant variation in the $\delta^{13}C$ of the C_4 plant material grown along the transect. Using the mean long-term discrimination values calculated on plant material grown under elevated CO_2 atmosphere having $\delta^{13}C$ close to the "normal" atmospheric value, it was then possible to reconstruct mean $\delta^{13}C$ of the source air along the transect using a mass balance approach to calculate average long-term day-time atmospheric CO_2 concentrations at the positions occupied by the plants along the transect (Bettarini et al., 1995).

Plantago major

Seed of two populations of *P. major* L. was collected from plants growing along a natural gradient of CO_2 concentrations at a CO_2 spring (Baccaiano) located near Siena in central Italy. This site, only recently discovered, comprises a large (10 cm) vent at the head of a small stream in a shallow valley. The vegetation at the site is typical of a Mediterranean spring-fed mire and the native vegetation predominantly consists of *Carex flacca* Shreber, *Juncus articulatus* L., *Phleum pratense* L., *Plantago major* L. and *Agrostis stolonifera* L., with *Phragmites australis* (Cav.) Trin ex Stoedel dominating the area immediately adjacent to the stream.

Seed of *P. major* originating 4 m and 40 m from the CO_2 vent was collected in 1994 at Baccaiano. Carbon dioxide measurements are restricted at this time to preliminary measurements made across the vent-site using Draeger diffusion tubes (Sestak, Catsky & Jarvis, 1971). These tubes enable accurate determinations of mean carbon dioxide concentration over a given period (maximum 8 h). For the purpose of this study, measurements of mean atmospheric CO_2 concentration were integrated over 4 h periods and then corrected for air temperature and pressure. As no CO_2 measurements were available at the time seed was collected at this site, experiments were also performed on an additional control population (originating from 350 µmol mol^{-1} CO_2) collected from a comparable climatic region (Anavriti, Spain).

Plant culture

Seed of two populations of *A. canina* and two populations of *P. major* (and a further control population collected from Anavriti, Spain), collected from sites exposed to contrasting atmospheric CO_2 concentrations was sown in a homogeneous peat-free compost (1:1 soil:sand fertilized to the equivalent of a standard nursery compost [John Innes no.2]) in pots of 1.14 dm^3 volume. At the early seedling stage plants were thinned to one plant pot^{-1}. Plants were grown from seed in duplicate controlled environment chambers maintained at target atmospheric CO_2 concentrations of 350 or 700 µmol mol^{-1} using an automated system enabling precise control of the atmospheric CO_2 concentration. These chambers comprise part of the University of Newcastle's phytotron and are described in detail elsewhere (Barnes *et al.*, 1995). Individual chambers were illuminated by metal-halide lamps providing a photon flux density (PFD) at leaf canopy height of 300 µmol m^{-2} s^{-1} as a 13.5 h photoperiod, extended to 14 h by fluorescent tubes positioned high above the chambers. Air temperature in the chambers was continuously monitored and was controlled on a dynamic basis from a midday maximum of 24±1°C to a night-time minimum of 14±1°C. Relative humidity inside the chambers was maintained at 65±5% and plants were watered to field capacity daily.

Growth and dry matter partitioning

Four pots, one from each of four fully randomized blocks within each chamber, were harvested at intervals over the course of the experimental period (48 and 98 days respectively for *P. major* and *A. canina*). Plants were removed from the pots in which they were grown, the roots washed and separated from the shoot and component plant parts oven-dried at 65°C for 4d before determining their dry weight. Relative growth rates (RGR) were calculated following the formula:

$$ln(W_2)-ln(W_1)/(t_2-t_1)$$

where W is the mass of the component plant part and t is time in weeks (Hunt, 1990). Projected leaf area was recorded using a leaf area meter (Delta-T Devices, Burwell, Cambridge, England) and the oven-dry weight of these leaves recorded. Specific leaf area (SLA) was calculated as decribed by Hunt (1990).

Statistical analyses

Statistical analyses were performed using SPSS (SPSS Inc., Chicago, Illinois). All datasets were first checked for normal distribution, then the influence of chamber, CO_2 and population determined by analysis of variance (ANOVA). Because of some missing values data were subjected to linear models for unbalanced data with the residuals from these models examined for normality. As no significant chamber to chamber variation was found within individual CO_2 treatments data were reanalysed using a reduced multivariate ANOVA (MANOVA) model concentrating on those factors that were significant and populations of *P. major* that were considered to be directly comparable. Time-course data were first subjected to a MANOVA model incorporating the factor "time", then data at each of the individual measurement dates were subjected to MANOVA. Finally, significant differences between treatments and populations were determined using the least significant difference (LSD) calculated at the 5% level.

RESULTS

Site CO_2 measurements

Atmospheric CO_2 concentration measurements made at the two CO_2 springs are shown in Fig. 1. At Solfatara, reconstructed average long-term CO_2 concentrations ranged from 422 to 644 µmol mol^{-1} along a 160 metre transect running north-southeast away from the major sources of CO_2 at this site with CO_2 concentrations decreasing with increasing distance from the vent (see Fig. 1a). Shorter term (integrated over 4 h periods) preliminary CO_2 measurements made at Baccaiano were, as might be expected, subject to greater variation than reliable estimates of long-term integrated CO_2 concentrations made at Solfatara using carbon isotope techniques. However, these measurements did provide some indication of the contrasting daytime CO_2 concentrations at the locations within the vent area at which seed of *P. major* was collected. Despite considerable variation, a significant ($P<0.05$) correlation was found between atmospheric CO_2 concentration and distance from the vent (see Fig. 1b). However, further detailed measurements are required to establish a more accurate picture of the CO_2 concentrations at this site.

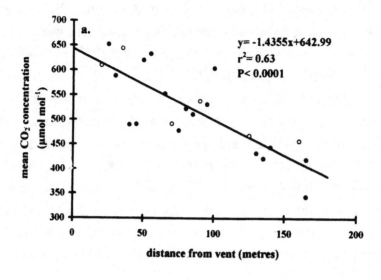

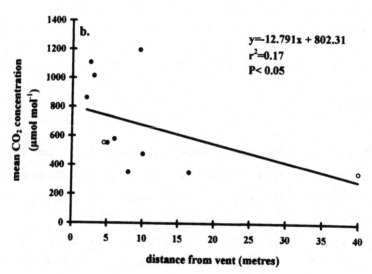

FIG. 1. Mean integrated CO_2 concentrations calculated from long-term carbon isotope discrimination along the transect at Solfatara (a) and calculated using integrated Draeger diffusion tubes at Baccaiano (b). Open symbols (O) represent points of seed collection.

Growth and dry matter partitioning

Fig. 2 shows the effects of controlled CO_2 enrichment on whole-plant biomass accumulation for two populations of *A. canina* and three populations of *P. major* originating from contrasting CO_2 concentrations. growth at elevated CO_2 [700 μmol mol^{-1}] increased (P<0.0001) biomass accumulation across populations in both species and whole plant dry weight was increased by 89% and 36% for *A. canina* and *P. major*, respectively, relative to ambient CO_2 grown plants. Consistent differences in biomass accumulation were found between populations (P<0.05) of both species. In *A. canina*, the high-CO_2 population produced a greater amount of dry matter than the control-CO_2 population when grown at both ambient and elevated CO_2, and these differences persisted throughout the duration of the experiment (Figure 2a). In *P. major*, the high-CO_2 population accumulated significantly (P<0.05) greater biomass when grown at elevated CO_2, but these differences were not maintained when plants were grown at ambient CO_2 (Fig. 2b).

CO$_2$ enrichment resulted in an initial stimulation in RGR in both species and the effect was significant (P<0.01) over the duration of the experiment (Fig. 3).

The initial stimulation in RGR induced by CO_2-enrichment appeared to be primarily responsible for the increase in biomass production and differences in population response (see Fig. 3). Relative to plants grown at ambient CO_2, the increase in RGR induced by elevated CO_2 after 22 days was 47% and 13% for high-CO_2 and control-CO_2 populations. Although there were significant (P<0.001) differences in original seed weight between populations of *P. major*, the magnitude of these differences was small (<20%). There was, however, a positive (r^2= 0.43, P<0.0005) correlation between initial RGR and seed weight at elevated CO_2, the larger seeds collected from plants adapted to high-CO_2 associated with a higher initial RGR. However, this relationship was not significant for plants grown at ambient CO_2.

Fig. 4 shows the initial RGRs of six populations of *A. canina* originating from the CO_2 transect at Solfatara.

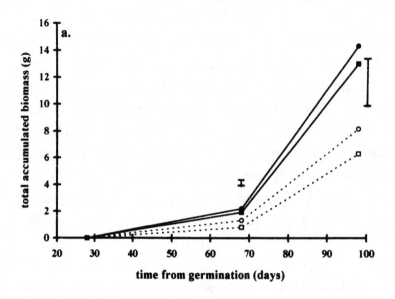

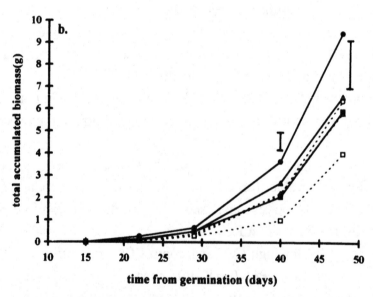

FIG. 2. Whole plant accumulated biomass (a) *A. canina* collected from high-CO_2 (O, ●) and control-CO_2 (□, ■) and (b) for *P. major* collected from high-CO_2 (O, ●), control-CO_2 (Δ, ▲) and Anavriti (□, ■); open symbols represent plants grown at 350 μmol CO_2 mol^{-1}; closed symbols represent plants grown at 700 μmol CO_2 mol^{-1}. Bars represent the least significant difference ($P<0.05$).

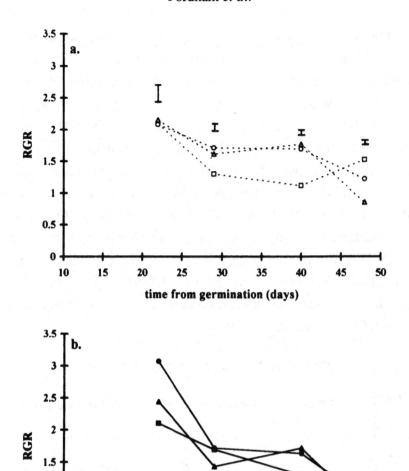

FIG. 3. Whole plant Relative Growth Rate (RGR) (a) *P. major* grown at 350 µmol CO_2 mol^{-1} (b) grown at 700 µmol CO_2 mol^{-1}. Populations; high-CO_2 (O, ●), control-CO_2 (Δ, ▲) and Anavriti (□, ■). Bars represent the least significant difference ($P<0.05$).

A significant linear relationship between initial plant RGR and the long-term average CO_2 concentration at the source of the plant material was found, with plants grown from seed collected at higher CO_2 concentrations exhibiting higher initial RGRs than plants grown from seed originating from lower CO_2 concentrations at both ambient and elevated CO_2. Original seed weight varied by only ≈10% between populations, but there was a highly significant (P<0.001) relationship between seed weight and long-term average CO_2 concentration at the source of the material (Fig. 5).

There was, however, a loose relationship between seed weight and initial RGR for plants grown at elevated CO_2 (P<0.001) and the relationship was not significant for plants grown at ambient CO_2.

The increase in biomass production induced by CO_2 enrichment was associated with effects on root:shoot (R:S) dry matter partitioning in *A. canina* (Fig. 6), with the R:S ratio significantly (P<0.01) increased by CO_2-enrichment in both populations indicating a relatively greater allocation of carbon to the root. MANOVA also revealed a significant (P<0.05) population x treatment interaction, plants originating from high-CO_2 allocating proportionately less carbon to the root under CO_2 enriched conditions than plants originating from near-ambient CO_2 concentrations (Fig. 6).

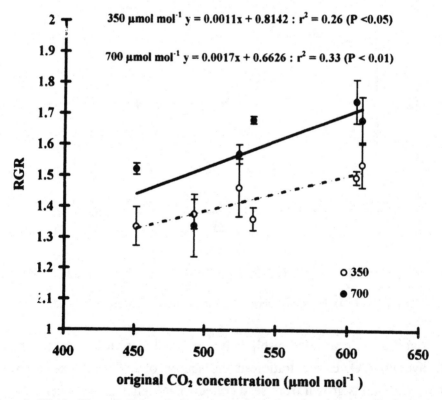

FIG. 4. Whole Plant Relative Growth Rate (RGR) of *A. canina* originating from contrasting CO_2 concentrations. Plants were grown at 350 (O) and 700 µmol CO_2 mol^{-1} (●).

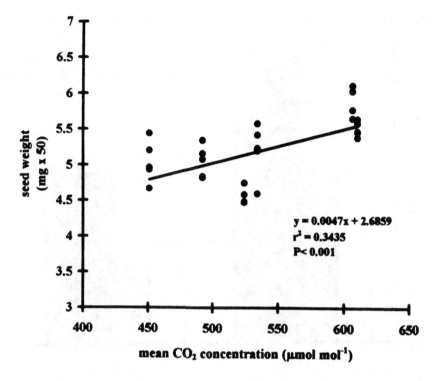

FIG. 5. Mean seed weights (n=50) of *A. canina* plotted against mean CO_2 concentration at the site of collection.

Changes in the R:S ratio induced by elevated CO_2 in *P. major* (data not shown) were not as consistent as those observed in *A. canina*, but there was evidence of a contrasting response; CO_2-enrichment resulting in a significant (P<0.05) decrease in R:S ratio in *P. major* indicating an increase in the relative partitioning of dry matter to the shoot. There were, however, no inherent differences in R:S ratio between populations of both species. Fig. 7 shows changes in SLA over the duration of the experiment for *P. major* originating from contrasting atmospheric CO_2 concentrations. Elevated CO_2 caused a significant (P<0.001) decrease in SLA in both populations, with the high-CO_2 population exhibiting a consistently greater reduction in SLA under CO_2-enriched conditions. Differences in SLA between populations were, however, only on the borderlines of statistical significance (pop P<0.06). Specific leaf area was also determined for *A. canina* after 90 days' exposure to ambient and elevated CO_2. These data revealed a similar CO_2-induced decrease in SLA (382 to 340 and 344 to

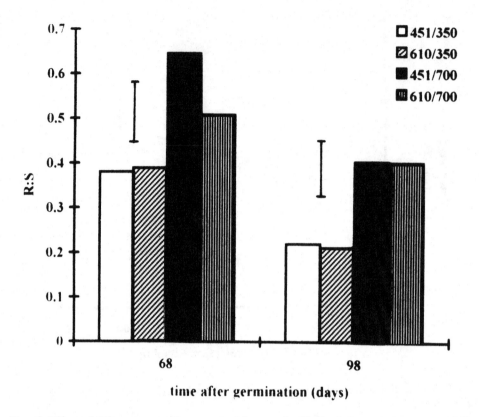

FIG. 6. Effect of CO$_2$ concentration on root/shoot ratio (R:S) of *A. canina* grown at 350 and 700 µmol CO$_2$ mol^{-1} for populations collected from control-CO$_2$ (451) and high-CO$_2$ (610); legend represents "source CO$_2$ / growth CO$_2$". Bars represent the least significant difference (P<0.05).

317 cm^2 g^{-1} for control-CO$_2$ and high-CO$_2$ populations, respectively) which represented a 10% decrease in SLA across populations.

Effects, as with *P. major*, were however only on the borderlines of significance (P<0.1), but the consistent differences between populations of both species indicated that plants originating from high-CO$_2$ concentrations tended to exhibit inherently lower SLA

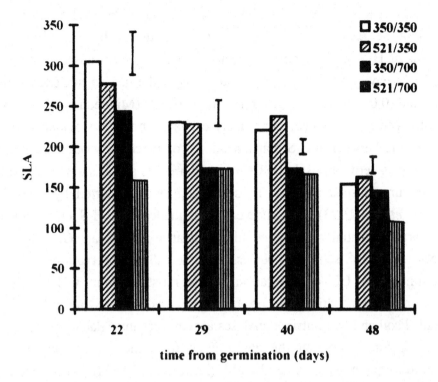

FIG. 7. Effect of CO$_2$ concentration on specific leaf area (SLA) of *P. major* grown at 350 and 700 µmol CO$_2$ mol^{-1} for populations collected at Baccaiano from control-CO$_2$ (350) and high-CO$_2$ (521); legend represents "source CO$_2$ / growth CO$_2$". Bars represent the least significant difference (P<0.05).

DISCUSSION

Site potential

Carbon dioxide springs offer the rare opportunity to examine plants and plant communities that have been exposed to naturally elevated CO$_2$ concentrations over evolutionary time-scales or periods long enough to allow "full acclimation". However, it is of clear importance that such sites should be fully characterized to enable phenotypic and genotypic responses to CO$_2$ to be distinguished from possible confounding effects. Particular attention has thus been paid to characterizing the physical and chemical environment at several CO$_2$ springs in central Italy, including Solfatara

(Miglietta & Raschi, 1993; Miglietta *et al.*, 1993a, 1993b; Selvi, 1996). At Solfatara the gas emitted by the vents is contaminated by H_2S. However, analysis of H_2S in air samples collected along the CO_2 transect shown in Figure 1a have revealed that concentrations of the gas do not exceed the threshold (0.03 μmol mol[-1]) for phytotoxic effects (Miglietta *et al.*, 1993a, 1993b). Moreover, biochemical analyses have shown no evidence of a relationship between leaf S concentrations, or indeed any other elements, and the long-term atmospheric CO_2 concentrations at this site (Fordham & Barnes, unpublished results). These findings are supported by those of Badiani *et al.* (1993) who showed no evidence that H_2S emissions at Solfatara substantially altered cellular S balance in soybean (*Glycine max* L. Merr.), an H_2S-sensitive species. Together, these observations indicate that atmospheric CO_2 concentrations are the most important variable along the transect from which seed of *A. canina* was collected for this study. Detailed soil and leaf nutrient analyses are, as yet, unavailable for the CO_2 spring at Baccaiano, but the gas emitted from the vent at this site is believed to be uncontaminated by H_2S (A. Raschi & F. Miglietta, personal communication).

Effects of controlled CO_2 enrichment

Plants well-supplied with nutrients and grown in containers that did not limit root growth showed a pronounced response to controlled CO_2 enrichment with relative increases in biomass production of 89% and 36% in *A. canina* and *P. major*, respectively. The magnitude of the growth response in *A. canina* is greater than that documented for many C_3 species in response to a doubling of the current atmospheric CO_2 concentration (Kimball *et al.*, 1990), but is within the range that might be expected for genotypes of the "competitive" functional type derived from strategy theory (Grime, 1979) which have been shown to respond strongly to elevated CO_2 (Hunt *et al.*, 1991, 1993). *P. major* responded to elevated CO_2 in a fashion similar to that reported by den Hertog, Stuhlen & Lambers (1993), where a 30% increase in dry matter production was found within three weeks of exposure but RGR was only increased during the initial period of growth. Although biomass production was substantially increased by elevated CO_2, particularly in those plants originating from high CO_2 concentrations, there was only a transitory stimulation of RGR implying that differences in plant stature at the end of the experiment were largely determined by what happened during the initial stages of growth. The factors underlying the decrease in RGR with sustained CO_2 enrichment are not well understood (den Hertog *et al.*, 1993; Poorter, 1993), but it is worthy of note that diagnostic gas exchange measurements made toward the end of the present study revealed no evidence of a down-regulation of photosynthesis in either *A. canina* or *P. major* (Fordham & Barnes, unpublished results). One suggestion is that RGR decreases with time due to increased investment in support tissue and an increase in self-shading (Poorter, 1993).

The impact of elevated CO_2 on R:S dry matter partitioning seen particularly in *A. canina* in the present study is difficult to place in a wider context as variable responses to elevated CO_2 have been reported both within and between species (Hunt *et al.*, 1991). The apparent variation in R:S allocation under CO_2 enriched conditions may be related to differences in available rooting volume as the impact of CO_2 enrichment on patterns of carbon allocation has been directly correlated with pot size (e.g. Arp, 1991; McConnaughay, Bernston & Bazzaz, 1993) and/or nutrient availability (McConnaughay *et al.*, 1993). In contrast to the variable responses observed in R:S partitioning, SLA has often been found to decline in response to CO_2 enrichment (Kimball *et al.*, 1990). Decreased SLA is generally associated with CO_2-induced increases in the accumulation of non-structural carbohydrates and secondary compounds in leaves (Lambers & Poorter, 1992) and/or changes in leaf thickness resulting from increases in the size and number of palisade cells (Thomas & Harvey, 1983).

CO₂ responsiveness of plants adapted to growth at contrasting atmospheric CO₂ concentrations

Growth analyses revealed significant differences in growth and dry matter partitioning in populations of both species originating from contrasting long-term atmospheric CO_2 concentrations. Plants adapted to growth at high CO_2 produced greater biomass and exhibited higher initial RGRs than plants originating from lower CO_2 concentrations. It is unclear at this stage whether the observed differences in growth potential represent the basis (i.e. phenotypic adaptation) or the by-product (i.e. genetic adaptation) of long-term adaptation to high CO_2. However, faster-growing plants would generally be expected to outcompete their slower-growing counterparts promoting eventual genetic changes within populations (Strain, 1987, 1991), although the extent to which numerous other factors (i.e. seed size, germination time, herbivory, nutrient acquisition, competition, reproductive success) may influence plant growth potential and mortality in the field is uncertain. For example, the production of larger seeds generally gives rise to better competitors at least during the early stages of growth (Stanton, 1984). Furthermore, plants originating from larger seeds have been shown to respond more strongly to elevated CO_2 (Bazzaz & Maio, 1993). This may reflect the fact that rapidly growing seedlings are considered to be 'source-limited' (Wardlaw, 1990) and may therefore have an increased capacity to utilize the additional carbohydrate resulting from growth at elevated CO_2 (Stitt, 1991). In the present study there was less than a 10% difference in original seed weight between *A. canina* populations, and less than a 20% difference in *P. major*. Yet, a positive correlation was found between seed weight and the long-term atmospheric CO_2 concentrations at the site of collection, and there was also a relationship between seed weight and initial RGR for plants grown at elevated CO_2. However, despite the significance of this latter relationship, differences in seed weight could explain no more than 34% and 43% of the variation in initial RGR in *A. canina* and *P. major*, respectively. Further work is needed to establish the genetic basis of the differences in seed weight, as this parameter is in part determined by maternal genotype as well as adult stature (Stanton, 1984).

The factors underlying the intrinsically higher growth potential of plants adapted to a high-CO_2 environment remain to be investigated further. Studies conducted by Fordham & Barnes (unpublished results) indicate no intrinsic differences in the photosynthetic capacity of *A. canina* populations originating from contrasting atmospheric CO_2 concentrations when plants are grown under optimal conditions. Increased growth potential in the present study was, however, associated with differences in R:S partitioning. In *A. canina*, plants originating from high-CO_2 partitioned a relatively greater proportion of the available dry matter to the shoot relative to plants originating from ambient CO_2 - an effect presumably reflected in an increased investment in the number and/or area of leaves, and therefore photosynthetic area. Moreover, initial differences in size would be expected to be magnified by the exponential nature of growth (Poorter, 1993).

Considerable variation in CO_2 responsiveness is known to exist within natural plant communities. However, it remains to be shown whether rising concentrations of CO_2 will act on this variation to select superior individuals in their existing habitat, and therefore ultimately result in shifts in the genetic structure of plant populations. The present study has documented intrinsic differences in the growth potential of plants adapted to elevated CO_2 and those adapted to present-day ambient CO_2 concentrations, but it remains to be shown whether this will confer a selective advantage under field conditions. It seems plausible at this stage to explain much of the observed differences in growth potential in terms of phenotypic plasticity rather than evoking evolutionary change. However, further work is needed in this, as yet, neglected area of global change research.

ACKNOWLEDGEMENTS

We thank Prof. J. Ehleringer (Dept. Biol., Utah State University) for conducting carbon isotope analyses and the staff of Close House Field Station (University of Newcastle) for their technical assistance. This work was supported by an EU grant attached to the 3rd Environment Programme (EV5V-CT94-0432) employing MF, with additional funding provided by The Royal Society (JB), the University of Newcastle's Equipment Trust (JB), the UKDoE (JB) and the UK NERC TIGER Programme at the University of Reading (IB).

REFERENCES

Allen, L.H. (1990) Plant responses to rising carbon dioxide and potential interactions with air pollutants. *Journal of Environmental Quality,* **19**, 15-34.

Arp, W.J. (1991) Effects of source-sink relations on photosynthetic acclimation to elevated CO_2. *Plant, Cell and Environment,* **14**, 869-875.

Badiani, M., D'Annibale, A., Paolacci, A.R., Miglietta, F. & Raschi, A. (1993) The antioxidant status of soybean (*Glycine max*) leaves grown under natural CO_2 enrichment in the field. *Australian Journal of Plant Physiology,* **20**, 275-284.

Barnes, J.D., Ollerenshaw, J.H. & Whitfield, C.P. (1995) Effects of elevated CO_2 and/or O_3 on growth, development and physiology of wheat (*Triticum aestivum* L.). *Global Change Biology,* **1**, 129-142.

Bazzaz, F.A. & Maio, S.L. (1993) Successional status, seed size, and responses of tree seedlings to CO_2, light and nutrients. *Ecology,* **74**, 104-112.

Bettarini, I., Giuntoli, A., Miglietta, F. & Raschi, A. (1995) Using stable carbon isotope ratio of C4 plants to monitor mean effective atmospheric CO_2 concentrations. *Agricoltura Mediterranea,* **Special Volume**, 278-281.

Conway, T.J., Trans, P., Waterman, L.S., Thoning, K.W., Masarie, K.A., & Gammon, R.M. (1988) Atmospheric carbon dioxide measurements in the remote global troposphere, 1981-1984. *Tellus,* **40**, 81-115.

den Hertog, J., Stuhlen, I. & Lambers, H. (1993) Assimilation, respiration and allocation of carbon in *Plantago major* as affected by atmospheric CO_2 levels. A case study. In *CO2 and Biosphere* (eds. J. Rozema, H. Lambers, S.C. Van De Geijn & M.L. Cambridge), pp. 369-378. Kluwer Academic Publishers, Dordrecht, The Netherlands.

Drake, B.G. & Leadley, P.W. (1991) Canopy photosynthesis of crops and native plant communities exposed to long-term elevated CO_2. *Plant, Cell and Environment,* **14**, 853-860.

Duchi, V., Minissale, A. & Romani L (1985) Studio geochimico su acque e gas dell'area geotermica Lago di Vico-M. Cimini (Viterbo). *Atti Soc. Tosc. Sci. Nat. Mem. Serie A,* **92**, 237-254.

Eamus, D. & Jarvis, P.G. (1989) The direct effects of increase in the global atmospheric CO_2 concentration on natural and commercial temperate trees and forests. *Advances in Ecological Research,* **19**, 1-55.

Friend, A.D. & Woodward, F.I. (1990) Evolutionary and ecophysiological responses of mountain plants to the growing season environment. *Advances in Ecological Research,* **20**, 59-124.

Grime, J.P. (1979) Primary strategies in plants. *Transactions of the Botanical Society of Edinburgh,* **43**, 151-160.

Grulke, N.E., Riechers, G.H., Oechel, W.C., Hjelm, U. & Jaeger, C. (1990) Carbon balance in tussock tundra under ambient and elevated atmospheric CO_2. *Oecologia,*

83, 485-494.

Hunt, R. (1990) *Basic Growth Analysis*. Unwin Hyman Press, London.

Hunt, R., Hand, D.W., Hannah, M.A. & Neal, A.M. (1991) Response to CO_2 enrichment in 27 herbaceous species. *Functional Ecology*, **5**, 410-421.

Hunt, R., Hand, D.W., Hannah, M.A. & Neal, A.M. (1993) Further responses to CO_2 enrichment in British herbaceous species. *Functional Ecology*, **7**, 661-668.

Kimball, B.A., Rosenberg, N.J., Heichel, G.H., Stuber, C.W., Kissel, D.E. & Ernst, S. (eds.) (1990*) Impact of Carbon Dioxide, Trace Gases, and Climate Change on Global Agriculture*. American Society of Agronomy, Madison, USA.

Körner, Ch. & Diemer, M. (1994) Evidence that plants from high altitudes retain their greater photosynthetic efficiency under elevated CO_2. *Functional Ecology*, **8**, 58-68.

Lambers, H. & Poorter, H. (1992) Inherent variation in growth rate between higher plants: A search for physiological causes and ecological consequences. *Advances in Ecological Research*, **23**, 187-261.

Maxon-Smith, J.W. (1977) Selection for response to CO_2-enrichment in glasshouse lettuce. *Horticultural Research*, **17**, 15-22.

McConnaughay, K.D.M., Bernston, G.M. & Bazzaz, F.A. (1993) Limitations to CO_2-induced growth enhancement in pot studies. *Oecologia*, **94**, 550-557.

Miglietta, F. & Raschi, A. (1993) Studying the effect of elevated CO_2 in the open in a naturally enriched environment in Central Italy. In *CO_2 and Biosphere* (eds. J. Rozema, H. Lambers, S.C. Van De Geijn & M.L. Cambridge), pp. 391-400. Kluwer Academic Publishers, Dordrecht, The Netherlands.

Miglietta, F., Raschi, A., Bettarini, I., Resti, R. & Selvi, F. (1993a) Natural CO_2 springs in Italy: a resource for examining long-term response of vegetation to rising atmospheric CO_2 concentrations. *Plant, Cell and Environment*, **16**, 873-878.

Miglietta, F., Raschi, A., Resti, R. & Badiani, M. (1993b) Growth and onto-morphogenesis of soybean (*Glycine max* L. Merr.) in an open, naturally CO_2-enriched environment. *Plant, Cell and Environment*, **16**, 909-918.

Petterson, R. & McDonald, J.S. (1994) Effects of nitrogen supply on the acclimation of photosynthesis to elevated CO_2. *Photosynthesis Research*, **39**, 389-400.

Poorter, H. (1993) Interspecific variation in the growth response of plants to an elevated ambient CO_2 concentration. In *CO_2 and Biosphere* (eds. J. Rozema, H. Lambers, S.C. Van De Geijn & M.L. Cambridge), pp. 77-98. Kluwer Academic Publishers, Dordrecht, The Netherlands.

Selvi, F. (1994) *Agrostis canina* L. ssp. *montelucci* Selvi (Poacee). *Webbia*, **49**, 1-9,

Selvi, F. (1996) Acidophilic grass community of CO_2-springs in central Italy: composition, structure and ecology. In this volume.

Sestak, Z., Catsky, J. & Jarvis, P.G. (1971) *Plant Photosynthetic Production. Manual of Methods*. W. Junk Publ. Co., The Hague.

Sonesson, M., Gehrke, C. & Tjus, M. (1992) CO_2 environment, microclimate and photosynthetic characteristics of the moss *Hylocomium splendens* in a subarctic habitat. *Oecologia*, **92**, 23-29.

Stanton, M.L. (1984) Size variation in wild radish: effect of seed size on components of seedling and adult fitness. *Ecology*, **65**, 1105-1112.

Stitt, M. (1991) Rising CO_2 levels and their potential significance for carbon flow in photosynthetic cells. *Plant, Cell and Environment*, **14**, 741-762.

Strain, B.R. (1987) Direct effects of increasing atmospheric CO_2 on plants and ecosystems. *Trends in Ecology and Evolution*, **2**, 18-21.

Strain, B.R. (1991) Possible genetic effects of continually increasing atmospheric CO_2. In *Ecological Genetics and Air Pollution*. (eds. G.E. Taylor, L.F. Pitelka & M.T. Clegg), pp. 237-244. Springer-Verlag, Berlin.

Strain, B.R. & Cure, J.D. (1986) Direct effects of atmospheric CO_2-enrichment on plants and ecosystems: bibliography with abstracts. Report ORNL/CDIC-13.

Thomas, J.F. & Harvey, C.N. (1983) Leaf anatomy of four species grown under continuous CO_2 enrichment. *Botanical Gazette*, **144**, 303-309.

Wardlaw, I.F. (1990) The control of carbon partitioning in plants. *New Phytologist*, **116**, 341-381.

Wulff, A.R. & Alexander, H.M. (1985) Intraspecific variation in the response to CO_2 enrichment in seeds and seedlings of *Plantago lanceolata*. *Oecologia*, **66**, 458-460.

Stomatal numbers in holm oak (*Quercus ilex* L.) leaves grown in naturally and artificially CO$_2$-enriched environments

E. PAOLETTI[1], F. MIGLIETTA[2], A. RASCHI[2], F. MANES[3]
AND P. GROSSONI[4]

[1]*C.S.P.S.L.M.-C.N.R., P.le delle Cascine 28, I-50144, Firenze, Italy.*
[2]*I.A.T.A.- C.N.R., P.le delle Cascine 18, I-50144, Firenze, Italy.*
[3]*Dipartimento Biologia Vegetale, Università "La Sapienza", P.le Aldo Moro 7, I-00185 Roma, Italy.*
[4]*Laboratorio Botanica Forestale, Dipartimento Biologia Vegetale, Università di Firenze, P.le delle Cascine 28, I-50144, Firenze, Italy.*

SUMMARY

The objects of the present study were:
1) to investigate the stomatal morphology of leaves of holm oak trees grown in a naturally CO$_2$-enriched environment;
2) to compare it with the stomatal density of leaves of holm oak seedlings grown in an artificially CO$_2$-enriched environment.

Among the stomatal morphology parameters we analysed, the only significant alteration we observed in trees which had grown by the CO$_2$ spring was a reduction in stomatal density.

The rather special response of one tree might be related to the rock on which it stands, which may be causing the tree severe water stress. However, the reduction did not increase significantly as the concentrations of CO$_2$ increased.

The experiment in artificially CO$_2$-enriched atmospheres indicated that the stomatal density of holm oak leaves was reduced by an increase in CO$_2$. The results are discussed with an analysis of herbarium holm oak leaves.

INTRODUCTION

The effect on stomatal density of the increase of atmospheric CO_2 is not yet fully clarified, although some experiments exposing plants to elevated CO_2 regimes in small scale chambers and field enclosures indicate that stomatal density decreases with increasing CO_2 concentrations (Madsen, 1973; O'Leary & Knetch, 1981; Thomas & Harvey, 1983; Imai, Coleman & Yanagisawa, 1984; Woodward, 1987), despite contradictory responses which were sometimes observed.

Criticisms have been levelled at this kind of experiment because of poor balance between energy supply and water loss from leaves (Morison, 1987). Conversely, natural vegetation is well coupled with the atmosphere, in terms of both its energy budget and its environmental feedback mechanisms.

A previous project (Paoletti & Gellini, 1993) studied the changes in stomatal density of *Quercus ilex* L. (holm oak) herbarium specimens over the last 200 years. The objects of the present study were 1) to investigate the stomatal morphology of leaves of holm oak trees grown in a naturally CO_2-enriched environment, the CO_2 spring of Bossoleto at Rapolano Terme (near Siena in central Italy); 2) to determine the stomatal density of leaves of holm oak seedlings grown in an artificially CO_2-enriched environment.

MATERIALS AND METHODS

Naturally CO_2-enriched environment

The area of the Bossoleto (Miglietta *et al.*, 1993) consists of a small enclosed doline, about 90 m wide and 7 m deep, at the bottom of which there are natural gas emissions rich in CO_2. The topographical conformation of the site is such that the atmospheric concentration of CO_2 gradually decreases as one moves upwards through the doline.

The arboreal component of the natural vegetation consists of *Quercus pubescens* Willd., *Fraxinus ornus* L. and above all *Quercus ilex*, an evergreen broadleaf that has a high drought tolerance and is one of the most representative elements of the Mediterranean maquis flora, where it originates the main climax formation.

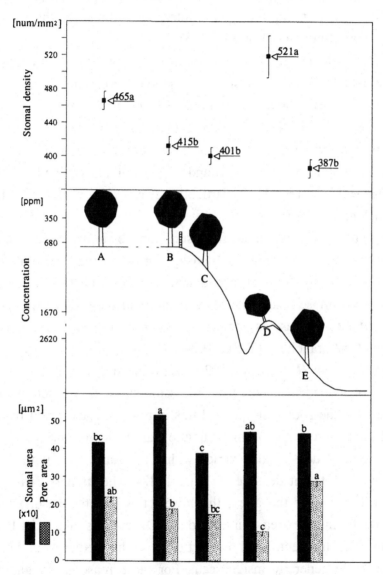

FIG. 1. Variations of stomatal morphology and CO_2 concentrations for holm oak trees at Bossoleto. Different small letters indicate significant differences (Tukey's test, P=0.05) between the sites of sampling (capital letters) in each measurement (stomatal density, stoma area, stoma pore area).

Fig. 1 shows the topographic profile of a hypothetical cross-section, where the distances are relative, but do reflect the trend of CO_2 concentrations. The CO_2 concentrations are daily mean values, measured with Draeger diffusion tubes or with a portable infra-red gas analyser, modified with a pump-fed gas diluter, according to the procedure illustrated by Miglietta & Raschi (1993).

The capital letters refer to the site of sampling. Site A is about 6 km from the actual CO_2 emission site, because in the zone immediately outside the doline - site B - the concentrations of CO_2 can at times, under special climate conditions, exceed the normal levels. Sites C, D and E are inside the doline and show progressively increasing concentrations of CO_2. Three trees were sampled for sites A, B and C, while in sites D and E only one tree was available. Tree D, in particular, grows on a rock right above a sink which has a CO_2 vent.

Leaf samples were gathered from sun branches in the lower section of the crown. Since the stomatal density varies during leaf ontogenesis (Ticha, 1982), only fully expanded leaves were collected, a preference being given to those nearer the apex of the branchlets, since the stomatal density tends to increase as one progresses from the basal leaves to those at the tip of the branchlets (Ticha, 1982).

Zalenski's law (Maximov, 1929) actually states: "The anatomical structure of individual leaves from the same shoot is a function of their distance from the root system", and this in practice means that the upper leaves have a more xeromorphic structure than the lower ones.

The stomatal density also varies within the same leaf, with different areas having different densities (Ticha, 1982). The highest densities are generally found near the apex, the lowest towards the base, while the density of the middle area is intermediate. Differences also exist between the outer edge, the centre and the part close to the median vein, but these differences are generally insignificant. For these reasons we gathered a minimum of four leaf segments from the central portion of each leaf sampled.

Each sample was air dried, then sputtered with carbon and gold (coater JEOL JEE 4B) and observed by means of a Philips 515 scanning electron microscope (SEM) at 15.0-15.2 kV. On each segment, avoiding the veins, we counted the number of stomata in eight random fields, for a total of about 0.5 mm^2 per leaf, 14 leaves per tree. The density and the two axes of the stomata and their pores (30 per leaf) were measured on SEM microphotographs. The surface areas of stomata and pores were then calculated, assuming that they were elliptical in shape.

The abaxial surface of holm oak leaves is covered by a pubescence of variable density. The use of SEM to measure stomatal density may sound surprising, but it is useful for observing the stomata in holm oak leaves. A slight mechanical abrasion with a sharp blade allowed us to remove the trichomes without causing serious damage to the stomata, but since the base of the removed trichomes is similar in shape and size to the stomata, a mere observation of the epidermal films with an optic microscope would not yield an accurate reading.

Therefore, it was necessary to employ SEM to differentiate stomata from trichome bases on abrated leaves. The use of SEM, however, did not allow for the measurement of stomatal index, that is, the ratio of stomatal to epidermal cells.

Artificially CO_2-enriched environment

Ten two-year-old holm oak seedlings were grown in controlled chambers (LABCO Mod. CC15) and maintained at one of two different CO_2 concentrations, 400 and 800 ppm, for 90 days. In both chambers, daytime temperature was 25°C and nightime temperature was 18°C for a photoperiod of 10 hours. The photosynthetic photon flux density at the height of the crown was 250 μmol m^{-2} s^{-1}.

The relative humidity of the air was 60%. Every day the seedlings were irrigated with water and once a week were fertilized with Hoagland solution. The mean air flow rate above the seedlings was 0.5 m s^{-1}. After 0, 55 and 90 days, ten recently sprouted but well sclerified leaves were gathered from each treatment group (one leaf per seedling).

After depilation by abrasion, a segment was removed from the median section of each leaf: this segment was air dried and metallized with carbon and gold.

Each leaf segment was observed under a Philips 515 scanning electron microscope: the number of stomata was counted in 10 randomly selected points, avoiding the veins, for a total of 0.4 mm^2 per leaf.

RESULTS

Except for tree D, the stomatal density of the samples gathered in the naturally CO_2-enriched environment was significantly lower than that of the site A trees (Fig. 1). However, the reduction did not increase progressively in a significant way as the concentrations of CO_2 increased. Neither the surface area of the stomata nor that of the pores appeared to respond in any significant way to the growing sites or to the increase of CO_2 (Fig. 1).

In the artificially CO_2-enriched chambers, stomatal density exhibited a dramatic progressive reduction at high CO_2 (Fig. 2). In the chamber at ambient CO_2 level a comparatively slight decrease in stomatal density compared to pre-test values was observed.

DISCUSSION

The stomatal density in the naturally CO_2-enriched environment was significantly lower than that of the site far from the CO_2 spring. The rather special response of tree D, despite the fact that it was growing directly over a vent producing high quantities of CO_2, might be related to the substrate on which it stands, which is made up of recently fractured rock. And this may be causing the tree severe water stress: it is well known that more arid conditions of plant growth usually cause higher stomatal density (Ticha, 1982). Also, a high stomatal density is a xeromorphic adaptation (Zalenski, 1904). We did not get a chance to measure the water availability during the period of leaf ontogenesis, so that this hypothesis still needs to be confirmed, although it appears to be further backed up by the reduced size of pore surface (Fig. 1); it has also been established that stoma length is shorter in xeric conditions (Gindel, 1968). However, among the

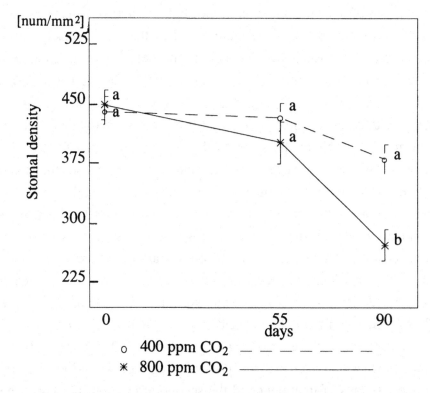

FIG. 2 Stomatal density of holm oak seedlings in artificially CO_2-enriched environment. Different letters indicate significant differences (Tukey's test, P=0.05) between the two CO_2 concentrations for each sampling day.

stomatal morphology parameters we analysed, the only significant alteration we observed in trees which had grown in the naturally CO_2-enriched environment was the reduction in stomatal density. Nevertheless, the reduction did not increase progressively in a significant way as the concentrations of CO_2 increased.

This might be explained in two ways: 1) the reduced stomatal density of the trees at Bossoleto may be an ecotypical adaptation and not related to the CO_2 emission; 2) the reducing effect of CO_2 on stomatal density may be counteracted by an increasing effect of other environmental factors, such as lower water availability.

The experiment in artificially CO_2-enriched atmospheres indicated that the stomatal density of holm oak leaves responds to an increase in CO_2. After 90 days of treatment, the stomatal density of the seedlings at high

CO_2 was about 65% of the initial one. The slight reduction observed in the ambient CO_2 treatment (-14%) was attributed to the light intensity inside the chambers, which was lower than that in the field where the time zero leaves had developed. Some authors (Schoch, 1972; Gay & Hurd, 1975; Wild & Wolf, 1980) have reported a positive relationship between irradiation and stomatal density. Note that time zero stomatal density (450 stomata mm^{-2}) was about the same as that of the ambient CO_2 trees at Rapolano Terme (site A), despite the fact that we compared seedlings and adult trees. The morphogenetic effect of CO_2 on stomatal density is, however, still not fully clarified: in the samples gathered after 55 days of the artificial treatment there were no significant differences in mean stomatal densities between the two CO_2 concentrations. The samples taken on day 55 certainly included leaves sprouted during treatment and already fully expanded. But these leaves were sprouted from buds that were already formed when the experiment began, i.e., buds which had formed when the seedlings were growing at ambient CO_2 concentrations. Seedlings showed two main growth flushes during the test period: the first began after 20 days of treatment and the second after about 60 days. The leaves sampled on day 90 were sprouted from buds formed during the test period, i.e., with the two different CO_2 concentrations. This would explain the different response of stomatal density on days 55 and 90, but imply that in the holm oak the codification of stoma initials occurs during leaf primordia formation. The hypothesis could justify some of the contradictory responses of stomatal density observed in simulation experiments of this type (see Apel, 1989). In any case, this result is an indication of the potential for acclimation in the new CO_2-enriched environment and stresses the importance of accurate sample gathering.

Woodward (1987) first demonstrated that the reduction of stomatal density could be seen in herbarium specimens of the leaves of several tree species collected over the past 200 years. He showed that as CO_2 levels have increased (as recorded from Antarctic ice cores), there has been a parallel drop in stomatal density. In a previous study Paoletti & Gellini (1993) measured the stomatal density in herbarium leaves of holm oak and of a species already studied by Woodward (1987), *Fagus sylvatica* L. (beech).

The selection and sampling of the material for this study on herbarium specimens was also done very carefully. Generally speaking, the same principles adopted for the sampling of the naturally CO_2-enriched site were followed in the herbarium as well, but with a few extra provisions. Only specimens bearing flowers or fruits were chosen as samples, since the reproductive shoots develop in full radiation, and it has been noted that sun leaves have a greater number of stomata than shade leaves (Solarova & Pospilova, 1988; Osborn & Taylor, 1990). All samples were gathered in two regions of central Italy (Tuscany and Latium) and all came from the Lauretum and Castanetum phytoclimatic horizons.

The results showed a 43% and 28% reduction in stomatal density over the last 200 years for beech and holm oak leaves, respectively, with a good statistical significance. The current mean value for the holm oak (475 stomata mm^{-2}) matched the mean levels of the ambient CO_2 trees growing at Rapolano Terme and those of the ambient CO_2 seedlings raised in the controlled chambers. De Lillis (1991) reported a lower value (266 stomata mm^{-2}) for holm oak leaves collected along the Latium coastline and analysed by impressions of the epidermis with cellulose acetate.

The reductions observed in the herbarium samples were comparable to the relative decrease in stomatal density of 40% measured by Woodward (1987) in seven tree and grass species, as well as with the 50% decrease observed by Beerling & Chaloner (1993b) in *Olea europaea* L. and with the 25% decrease recorded by Peñuelas & Matamala (1990) in 14 trees, shrubs and grasses in the Mediterranean region (but they did not include the holm oak).

Such a reduction in holm oak stomatal density, which as mentioned earlier also means a reduction in the tree's xeromorphic characteristics, appears to raise some questions about the ecological future of this species in the event that atmospheric CO_2 concentrations continue to increase. But, by analogy with the results of growth chamber experiments (Woodward & Bazzaz, 1988) and with the findings of our study, stomatal density may appear to respond to increases of CO_2, but the degree of the response may be related to the level of the existing concentration: higher responses of stomatal density were found during the first stages of CO_2 enrichment. In actual fact, in the naturally CO_2-enriched site, an increase of

CO_2 of more than 2600 ppm led to a mere 17% decrease in stomatal density, whereas the much lower CO_2 increase recorded from the preindustrial age to the present would have brought about a 28% reduction.

It is also possible that other factors contribute to reducing the effect of CO_2 increases on the stomatal density of trees in the naturally enriched site: factors such as the above-mentioned lack of water. According to the comments of Beerling & Chaloner (1993a) for herbarium material, the response of stomatal density shown by CO_2-spring vegetation may be the product of a combination of environmental factors. The experiment with artificially CO_2-enriched atmospheres appeared to suggest that holm oak stomatal density is reduced by high CO_2 levels. However, we cannot state that this is the only factor influencing the stomatal density reductions observed in natural conditions.

ACKNOWLEDGEMENT

The cost of the research has been partially covered by EU, Programme Environment, contract EV5V-CT93-0292.

REFERENCES

Apel, P. (1989) Influence of CO_2 on stomatal numbers. *Biologia Plantarum (Praha)*, **31**, 72-74.

Beerling, D.J. & Chaloner, W.G. (1993a) The impact of atmospheric CO_2 and temperature change on stomatal density: observations from *Quercus robur* lammas leaves. *Annals of Botany*, **71**, 231-235.

Beerling, D.J. & Chaloner, W.G. (1993b) Stomatal density responses of Egyptian *Olea europaea* L. leaves to CO_2 changes since 1327 BC. *Annals of Botany*, **71**, 431-435.

De Lillis, M. (1991) An ecomorphological study of the evergreen leaf. *Braun-Blanquetia*, **7**, 1-127.

Gay, A.P. & Hurd, R.G. (1975) The influence of light on stomatal density in the tomato. *New Phytologist*, **75**, 37-46.

Gindel, I. (1968) Dynamic modifications in alfalfa leaves growing in subtropical conditions. *Physiologia Plantarum*, **21**, 1287-1295.

Imai, K., Coleman, D.F. & Yanagisawa, T. (1984) Elevated atmospheric partial pressure of carbon dioxide and dry matter production of Cassava (*Manihot esculenta* Crantz). *Japanese Journal of Crop Science*, **53**, 175-184.

Madsen, E. (1973) Effect of CO_2-concentration on morphological, histological, cytological and physiological processes in tomato plants. *Acta Agric. Scand.*, **23**, 241-246.

Maximov, N.A. (1929) *The plant in relation to water. A study of the physiological basis of drought resistance.* George Allen and Unwin Ltd, London.

Miglietta, F. & Raschi, A. (1993) Studying the effect of elevated CO_2 in the open in a naturally enriched environment in Central Italy. *Vegetatio*, **104/105**, 391-400.

Miglietta, F., Raschi, A., Bettarini, I., Resti, R. & Selvi, F. (1993) Natural CO_2 springs in Italy: a resource for examining long-term response of vegetation to rising atmospheric CO_2 concentrations. *Plant, Cell and Environment*, **16**, 873-878.

Morison, J.I.L. (1987) Plant growth and CO_2 history. *Nature*, **327**, 560.

O'Leary, J.W. & Knetch, G.N. (1981) Elevated CO_2 concentration increases stomata numbers in *Phaseolus vulgaris* leaves. *Botanical Gazette*, **142**, 438-441.

Osborn, J.M. & Taylor, T.N. (1990) Morphological and ultrastructural studies of plant cuticular membranes. I. Sun and shade leaves of *Quercus velutina* (Fagaceae). *Botanical Gazette*, **151**, 465-476.

Paoletti, E. & Gellini, R. (1993) Stomatal density variation in beech and holm oak leaves collected over the last 200 years. *Acta Oecologica*, **14**, 173-178.

Peñuelas, J. & Matamala, R. (1990) Changes in N and S leaf content, stomatal density and specific leaf area of 14 plant species during the last three centuries of CO_2 increase. *Journal of Experimental Botany*, **41**, 1119-1124.

Schoch, P.G. (1972) Effect of shading on structural characteristics of the leaf and yield of fruit in *Capsicum annuum* L. *Journal of American Society Horticultural Science*, **97**, 461-464.

Solarova, J. & Pospilova, J. (1988) Stomatal frequencies in the adaxial and abaxial epidermis of primary leaves affected by growing irradiances. *Acta Univ. Carol.-Biol.*, **31**, 101-105.

Thomas, J.F. & Harvey, C.N. (1983) Leaf anatomy of four species grown under continuous CO_2 enrichment. *Botanical Gazette*, **144**, 303-309.

Ticha, I. (1982) Photosynthetic characteristics during ontogenesis of leaves. 7. Stomata density and sizes. *Photosynthetica*, **16**, 375-471.

Wild, A. & Wolf, G. (1980) The effect of different light intensities on the frequency and size of stomata, the size of cells, the number, size and chlorophyll content of chloroplasts in the mesophyll and the guard cells during the ontogeny of primary leaves of *Sinapis alba*. *Z. Pflanzenphysiol.*, **97**, 325-342.

Woodward, F.I. (1987) Stomatal numbers are sensitive to increases in CO_2 from pre-industrial levels. *Nature*, 127, 617-618.

Woodward, F.I. & Bazzaz, F.A. (1988) The response of stomatal density to CO_2 partial pressure. *Journal of Experimental Botany*, **39**, 1771-1781.

Zalenski, V. (1904) Materialy K Kolichestvennoiï anatomiï razlichnikh list'ev odnikh i tekh zhe rasteniï. *Mem Politekh. Inst. Kiev*, **4**, 1-203.

Effects of CO_2 on NH_4^+ assimilation by *Cyanidium caldarium*, an acidophilic hot springs and hot soils unicellular alga

C. RIGANO, V. DI MARTINO RIGANO, V. VONA, S. ESPOSITO
AND C. DI MARTINO

Dipartimento di Biologia Vegetale, Università di Napoli Federico II, Via Foria 223, 80139, Napoli, Italy.

SUMMARY

The effect of CO_2 and light on NH_4^+ assimilation by *Cyanidium caldarium*, a thermophilic acidophilic unicellular alga isolated in volcanic areas in Yellowstone National Park, USA, was investigated. N-sufficient cells assimilated NH_4^+ at a rate of 189 ± 4.76 µmol ml^{-1} packed cell volume (pcv) h^{-1}. Removal of CO_2 or darkening almost immediately prevented NH_4^+ assimilation. N-limited cells in light assimilated NH_4^+ at the rate of 498 ± 8.05 µmol ml^{-1} pcv h^{-1} in the presence of CO_2. In darkness they assimilated NH_4^+ at a rate of 303 ± 1.5 µmol ml^{-1} pcv h^{-1} in the presence of CO_2, which was as high as 60% of assimilation in the light, and at a similar rate in the absence of CO_2. However, after 40 min under the latter conditions, assimilation underwent a time dependent inhibition and ceased after 70 min; it was resumed upon resupply of CO_2. In the absence of CO_2 in light, NH_4^+ was assimilated at a considerably lower rate than in darkness, which supports the idea that, under CO_2-free conditions, a light-dependent inhibition of NH_4^+ assimilation occurred. These results are consistent with the contention that cells of *C. caldarium*, grown under excess NH_4^+, obtain carbon skeletons for NH_4^+ assimilation exclusively by photosynthetic reactions, thereby the light and CO_2 dependence. Cells grown under conditions of N-limitation possess the ability to obtain a consistent amount of additional carbon skeletons from mobilization of carbon reserves. This enables N-limited cells to assimilate NH_4^+ not only in light but also in darkness. In light, NH_4^+ assimilation by N-limited cells is the sum of both light and dark assimilation and thus, in these cells, NH_4^+ is assimilated at a higher rate than in N-sufficient cells. The fundamental role of dark-CO_2 fixation in dark or light NH_4^+ assimilation is also emphasized. It is also suggested that in *C. caldarium* CO_2 could indirectly affect regulatory mechanisms at the level of NH_4^+ uptake.

INTRODUCTION

There are numerous microorganisms, either heterotrophic or autotrophic, found in volcanic springs or fumaroles. These organisms from extreme environments are now studied extensively, either for the peculiarities of their physiology, or to select strains of extremophile microorganisms suitable for development or improvement of new biotechnologies (Aguilar, 1988).

Cyanidium caldarium is a thermophilic acidophilic unicellular alga which lives in hot acidic volcanic environments, namely fumaroles or hot springs in Java (Geitler, 1936), Japan (Hirose, 1950), USA (Brock, 1969; Doemel & Brock, 1971), Italy (Rigano & Conforti, 1967; De Luca, Taddei & Moretti, 1973), etc. It shares its habitat with another eukaryotic alga, *Galdieria sulphuraria* (Merola *et al.*, 1981). *C. caldarium* is a strict autotrophic organism. It is unable to use organic compounds for heterotrophic growth; it uses NH_4^+ or NO_3^- as its sole nitrogen source (Rigano, 1971). It can also use glutamate, but glutamate is a poor nitrogen source (Rigano, Aliotta & Violante, 1974). The ecology of this alga is very peculiar: it lives in a temperature range from 30°C to 55°C, and at very low pH values (between 1, or less, and 5). Springs (Doemel & Brock, 1971) or soils (Rigano & Conforti, 1967) where *Cyanidium* occurs are rich in N-NH_4^+, and fumaroles contain water vapour, CO_2, and H_2S.

Studies on mineral nitrogen nutrition in photosynthetic organisms have shown that assimilation of NH_4^+ by N-sufficient unicellular algae depends strictly on light and CO_2, supporting the idea that the necessary carbon skeletons are supplied exclusively by photosynthesis (Turpin, 1991). However, as observed in several unicellular algae, dependence on light does not occur in cells submitted to a period of nitrogen starvation (Syrett, 1956; Thacker & Syrett, 1972), or in cells grown in continuous cultures under conditions of nitrogen limitation (Di Martino Rigano *et al.*, 1986). In this paper, effects of CO_2 on NH_4^+ assimilation by *Cyanidium caldarium* are studied, in order to draw further elucidation concerning nutritional characteristics of photosynthetic organisms which live in natural environments where CO_2 and NH_4^+ are abundant.

MATERIALS AND METHODS

Cyanidium caldarium, strain 0206, (obtained from Prof. T.D. Brock, University of Wisconsin), was isolated in Yellowstone National Park (Wyoming) USA. For these experiments the alga was grown autotrophically at pH 1.9 and at 42 °C, either in batch culture, where all mineral nutrients were present in concentrations exceeding those needed for growth, or in a chemostat under conditions of NH_4^+ limitation.

The basal composition of the medium was the following: 2.2 mM KH_2PO_4; 0.18 mM $CaCl_2$; 1.21 mM $MgSO_4$; 20 mM H_2SO_4; 71 µM Fe, 0.31 µM Cu, 0.1 µM Mo, 9.1 µM Mn, 0.76 µM Zn and 46 µM B. The medium used for batch culture contained a final 10 mM $(NH_4)_2SO_4$. The medium used for N-limited chemostat culture contained a final 2 mM $(NH_4)_2SO_4$, and was added to the culture at a dilution rate of 0.20 day^{-1}. The cultures were maintained under conditions of continuous light (200 µmol photons m^{-2} s^{-1}, Philips TLD 30 W/55 fluorescent lamps), and were continuously flushed with air containing 5% CO_2. The culture apparatus was as previously reported (Rigano *et al.*,1981).

NH_4^+ uptake

Cells, collected and washed with N-free basal medium by low speed centrifugation (4000 x g), were resuspended at a density of 1.5-3 µl pcv (packed cell volume) ml^{-1} into medium containing NH_4^+ given as $(NH_4)_2 SO_4$. The cell suspensions were then kept at 42 °C in the light or darkness, with or without CO_2, as described in the text. To remove CO_2, air was flushed through 10 M NaOH before it reached the cell suspensions.

NH_4^+ was assayed continuously by monitoring its disappearance from the external medium with an ammonia-specific electrode (Di Martino Rigano *et al.*, 1986). The electrode was connected to an ionanalyser (Orion model EA 920) in concentration mode and connected to a serial printer. The decrease in NH_4^+ concentration of cell suspensions enabled measurement of NH_4^+ utilization rates.

Oxygen exchange

Respiration rate was measured in a BOD bottle by the aid of an Orion oxygen electrode connected to an EA 920 ionanalyser in O$_2$ mode and connected to a serial printer.

Packed cell volume determination

PCV was estimated by centrifuging a known aliquot of cell suspension in a haematocrit tube.

RESULTS

Rates of NH$_4^+$ assimilation by N-sufficient and N-limited cells of Cyanidium caldarium

N-sufficient cells of *C. caldarium* assimilated NH$_4^+$ at a mean value (± SE) of 189 ± 4.76 µmol ml^{-1} pcv h^{-1} when supplied with light and CO$_2$. In darkness assimilation was very low or undetectable (Table 1). N-limited chemostat cells of *C. caldarium* supplied with CO$_2$ in the light assimilated NH$_4^+$ at a rate of 498 ± 8.05 µmol ml^{-1} pcv h^{-1}, which was about 2.7-fold higher than the rate exhibited by N-sufficient cells under the same conditions.

N-limited cells also assimilated NH$_4^+$ in darkness, although at a lower rate of 303 ± 1.5 µmol ml^{-1} pcv h^{-1} (Table 1).

The difference between the rate of light and dark NH$_4^+$ assimilation by N-limited cells was 200 µmol ml^{-1} pcv h^{-1}, remarkably similar to the value of the rate of the solely light-dependent NH$_4^+$ assimilation occurring in N-sufficient cells.

Patterns of NH$_4^+$ assimilation

Fig. 1 shows that NH$_4^+$ assimilation by N-sufficient cells under light and CO$_2$ conditions was linear with time.

TABLE 1. Rates of NH_4^+ uptake by cell suspensions of N-sufficient or N-limited *C. caldarium* in light and darkness. Results are the average of 10 independent experiments for each condition. Cell suspensions were aerated with air enriched with 5 % CO_2.

Type of cell	NH_4^+ uptake (μmol NH_4^+ ml^{-1} pcv h^{-1} ± SE)	
	Light	Dark
N-sufficient	189 ± 4.76	0
N-limited	498 ± 8.05	303 ± 1.5

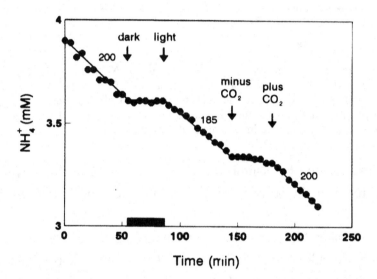

FIG. 1. Effect of light and CO_2 on NH_4^+ assimilation by N-sufficient cells of *Cyanidium caldarium*. The cell suspension was kept in the light at 42 °C and was flushed with air containing 5% CO_2. At the time indicated by the first arrow the cell suspension was darkened and then was illuminated (second arrow). At the time indicated by the third arrow CO_2 was removed from the sufflating air and then again supplied (fourth arrow). The numbers above the trace lines are rates of NH_4^+ uptake activity expressed as μmol NH_4^+ absorbed ml^{-1} pcv h^{-1}.

Inhibition of NH_4^+ assimilation upon darkening was immediate. Assimilation was also inhibited by removal of CO_2. Resumption of NH_4^+ assimilation upon resupply of light and of CO_2 was immediate.

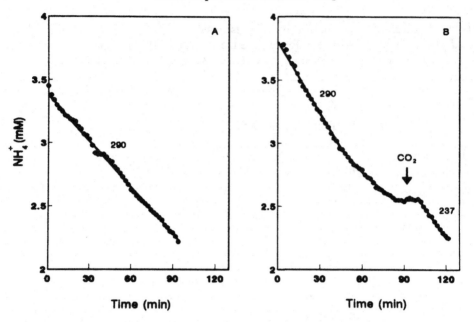

FIG. 2A, B. Effect of CO_2 deprivation on NH_4^+ uptake by N-limited cells of *Cyanidium caldarium* in darkness. The cell suspensions, kept at 42°C in darkness, were flushed with air containing 5% CO_2 (A) or with CO_2-free air (B). Cell densities: 2.5 µl pcv ml^{-1} in A; 4 µl pcv ml^{-1} in B. At the time indicated by the arrow the CO_2 (5%) was again added. Numbers above the trace lines same as described in Fig. 1.

NH_4^+ assimilation by N-limited cells in darkness was linear with time and occurred at a similar rate both in the presence and in the absence of CO_2 (Fig. 2).

However, after 40 min incubation in the presence of CO_2, NH_4^+ assimilation was still linear (Fig. 2A). Under CO_2-free conditions it decreased with time and had almost ceased after about 70 min (Fig. 2B). Addition of CO_2 restored 65% of the initial rate of NH_4^+ assimilation in darkness.

Under light conditions, and after only 6 min of CO_2 deprivation, there occurred a 75% inhibition of NH_4^+ assimilation by N-limited cells (Fig. 3).

Addition of CO_2 restored NH_4^+ assimilation completely. In contrast, under dark conditions, CO_2 deprivation did not produce inhibition of NH_4^+ assimilation after 6 min (see also Fig. 2B).

Effect of NH_4^+ on dark respiration

N-sufficient cells and N-limited cells of *C. caldarium* resuspended under N-free conditions showed similar rates of dark respiratory O_2 consumption of

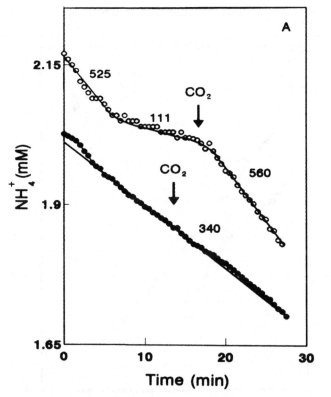

FIG. 3. Effect of CO_2 deprivation on NH_4^+ uptake by N-limited cells of *Cyanidium caldarium* kept in the light (O-O) and in darkness (●-●) and gassed with CO_2-free air. At the time indicated by the arrows CO_2 (5%) was added again. Numbers above the trace lines same as described in Fig 1.

185 and 200 μmol ml^{-1} pcv h^{-1}, respectively. Addition of NH_4^+ resulted in up to a 2-fold stimulation in dark O_2 consumption by N-limited cells and had no effect on O_2 consumption by N-sufficient cells (Fig. 4).

DISCUSSION

Our data show that N-sufficient cells of *Cyanidium caldarium* assimilate NH_4^+ only when supplied with both light and CO_2, which agrees withthe hypothesis that, in this type of cell, assimilation requires active photosynthesis. This finding is consistent with reports in spinach and in other algae showing that the Calvin cycle triose phosphate can be a primary source of carbon for NH_4^+ assimilation and amino acid synthesis (Kanazawa, Kirk & Bassham, 1970; Kanazawa *et al.*, 1972; Larsen *et al,*.

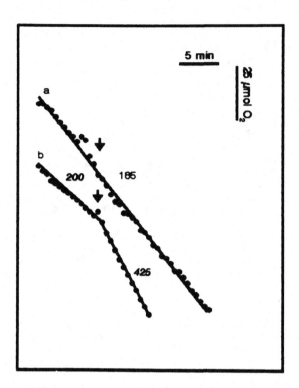

Fig. 4. Effect of NH_4^+ on dark respiratory oxygen consumption by N-sufficient (a) and by N-limited chemostat (b) cells of *C. caldarium*. The cells (1.5 µl ml⁻¹ pcv in a and 1µl ml⁻¹ pcv in b) were resuspended at 42 °C in darkness in the same basal medium used for growth; 1 mM NH_4^+ was added at the time indicated by the arrows. The numbers above the curve denote µmol O_2 ml⁻¹ pcv h⁻¹.

1981; Turpin, 1991). The strict dependence on photosynthesis supports the idea that N-sufficient cells obtain all of the carbon skeletons needed for nitrogen assimilation in this manner.

N-limited cells of *C. caldarium* assimilated NH_4^+ in the light at a higher rate than N-sufficient cells and were also able to assimilate NH_4^+ in darkness. This dark NH_4^+ assimilation fits in with the idea that N-limited cells are able to draw carbon skeletons for NH_4^+ assimilation from carbon reserves (Turpin, 1991). Noteworthy is that, as shown with N-limited cells, NH_4^+ stimulated respiration of *Cyanidium*, an event able to meet the major need of carbon skeletons and ATP required for reactions of NH_4^+ incorporation and amino acid synthesis. This is consistent with the finding of Miyachi & Miyachi (1985) that addition of NH_4^+ in *Chlorella* stimulated breakdown of starch.

N-limited *Cyanidium* assimilated NH_4^+ at a lower rate in darkness than in light, supporting the notion that, in the light, N-limited cells are able to operate concomitantly light-dependent plus dark NH_4^+ assimilation. Light-dependent NH_4^+ assimilation by N- limited cells occurred at a rate similar to that of light-assimilation exhibited by N-sufficient cells. Therefore, there exists a well defined amount of carbon compounds that either N-sufficient or N-limited cells can derive from the Calvin cycle to use for NH_4^+ assimilation, which is equivalent in both types of cell. It follows that the 2.7-fold increase in NH_4^+ assimilation rate, exhibited under light conditions by N-limited cells, is exclusively due to the fact that, as referred to above, these cells have developed the ability to use carbon reserves as a source of carbon skeletons for NH_4^+ assimilation, in addition to those derived from the Calvin cycle. The fact that N-limited *C. caldarium* assimilated NH_4^+ in the light at a velocity which was the sum of light-dependent activity plus dark-activity, suggests that separately Calvin cycle or mobilization of carbon reserves are inadequate to supply N-limited cells with the carbon needed for full NH_4^+ assimilation. In these cells NH_4^+ incorporation rate may be limited by the availability of carbon skeletons.

In the dark under CO_2 conditions, NH_4^+ assimilation by N-limited cells was linear with time, whereas under CO_2-free conditions it was linear for 40 min, after which it exhibited a time-dependent decrease, perhaps because in the absence of CO_2, the dark CO_2 fixation was prevented. This supports that dark CO_2 fixation plays a fundamental role in NH_4^+ assimilation for anaplerotic reactions through which intermediate compounds of Krebs cycle used for NH_4^+ incorporation and amino acid synthesis are restored (Ohmori *et al.*, 1984). It is possible that initially endogenously produced CO_2 through increased respiration was able to sustain dark CO_2 fixation. The fact that in the light, under CO_2-free air, N-limited cells exhibited an almost 75% inhibition of NH_4^+ assimilation in less than 10 min, suggests that the photosynthetic carboxylation will compete with dark carboxylation for the limited amount of CO_2 produced by respiration. Hence the production of carbon precursors through dark CO_2 fixation for amino acid biosynthesis will be slower in the light than in the dark and, as a consequence, NH_4^+ assimilation will also be slower. An alternative hypothesis is offered by the observation that in a mutant of *Chlamydomonas*, devoid of ribulose bisphosphate carboxylase, and thus unable to assimilate CO_2, the rate of respiratory CO_2 evolution in the dark was diminished by 60% to 85% in the light, and that darkness or addition of 3-(3,4-dichlorophenyl)-1,1-dimethylurea (DCMU), an inhibitor of photosynthetic electron transport, fully reversed the effect of light (Gans & Rebeille, 1988). Similarly, one may suppose that the light decline of NH_4^+ assimilation occurring in *Cyanidium* in the absence of exogenous CO_2 could be due to a diminished evolution of respiratory CO_2. In autotrophic organisms this would be a safety mechanism, which would enable them to economize on carbon reserves under conditions where these reserves cannot be restored through photosynthesis because of the absence of an exogenous CO_2 source.

The inability of N-sufficient cells of *C. caldarium* to assimilate NH_4^+ when CO_2 is removed does not necessarily indicate a strict dependence of NH_4^+ assimilation on photosynthesis. It is possible also that availability of carbon compounds from recent photosynthesis favours the further metabolism of nitrogen substance(s) which could act as a regulatory molecule in a short-term control of NH_4^+ uptake (Thacker & Syrett, 1972). This hypothesis is consistent with findings in cyanobacteria, where molecules derived from NO_3^- assimilation acted as negative effectors triggering a regulatory mechanism through which NO_3^- uptake was prevented. Light conditions, through a supply of carbon skeletons effective in the removal of the inhibitors, will restore NO_3^- uptake activity (Lara & Romero, 1986; Lara *et al.*, 1987; Romero *et al.*, 1987).

The dependence on CO_2 of NH_4^+ utilization suggests that natural spots rich both in CO_2 and NH_4^+ are particularly suitable for growth and development of those unicellular photosynthetic organisms which became adapted to live in these environments.

ACKNOWLEDGEMENTS

We are grateful to Prof. T.D. Brock for the generous gift of a pure strain, 0206, of *Cyanidium caldarium*. Research supported by National Research Council of Italy, Special Project RAISA, sub project N. 2 Paper N. 2474.

REFERENCES

Aguilar, A. (1988) European laboratory without walls on extremophiles. In *Microbiology of extreme environments and its potential for biotechnology* (eds. M.S. Da Costa, J.C. Duarte & R.A.D Williams), pp. 1-5. Proc. Fed. European Microbiological Society. Elsevier, London and New York.

Brock, T.D. (1969) Microbial growth under extreme conditions. *Symposia of the Society for General Microbiology. Microbial growth.*, **XIX**, 15-41.

De Luca, P., Taddei, R. & Moretti, A. (1973) Nuove stazioni di "*Cyanidium caldarium*" nell'Italia meridionale ed in Sicilia. *Delpinoa*, **14-15**, 49-60.

Di Martino Rigano, V., Vona, V., Di Martino, C. & Rigano, C. (1986) Effect of darkness and CO_2 starvation on NH_4^+ and NO_3^- assimilation in the unicellular alga *Cyanidium caldarium*. *Physiologia Plantarum*, **68**, 34-38.

Doemel, W.N. & Brock, T.D. (1971) The physiological ecology of *Cyanidium caldarium*. *Journal of General Microbiology*, **67**, 17-32.

Gans, P. & Rebeille, F. (1988) Light inhibition of mitochondrial respiration in a mutant of *Chlamydomonas reinhardtii* devoid of ribulose-1,5-bisphosphate carboxylase/oxygenase activity. *Archives of Biochemistry and Biophysics*, **260**, 109-117.

Geitler, L. (1936) Die Cyanophyceen der Deutschen Limnologischen Sunda Expedition. *Archives f. Hydrobiologie*, **14**, 388-391.

Hirose, H. (1950) Studies on *Cyanidium caldarium* (Tilden) Geitler, with special reference to its ecology and distribution in Japan. *Journal of Japanese Botany*, **25**, 179-184.

Kanazawa, T., Kanazawa, K., Kirk, M.R. & Bassham, J.A. (1972) Regulatory effects of ammonia on carbon metabolism in *Chlorella pyrenoidosa* during photosynthesis and respiration. *Biochimica et Biophysica Acta*, **256**, 656-669.

Kanazawa, T., Kirk, M.R. & Bassham, J.A. (1970) Regulatory effects of ammonia on carbon metabolism in photosynthesizing *Chlorella pyrenoidosa*. *Biochimica et Biophysica Acta*, **205**, 401-408.

Lara, C. & Romero, J.M. (1986) Distinctive light and CO_2-fixation requirement of NO_3^- and NH_4^+ utilization by the cyanobacterium *Anacystis nidulans*. *Plant Physiology*, **81**, 686-688.

Lara, C., Romero, J.M., Coronil, T. & Guerrero, M.G. (1987) Interaction between photosynthetic NO_3^- assimilation and CO_2 fixation in Cyanobacteria. In *Inorganic nitrogen metabolism* (eds. W. Ullrich *et al.*), pp. 45-52. Springer-Verlag, Berlin Heidelberg.

Larsen, P.O., Cornwell, K.L., Gee, S.L. & Bassham, J.A. (1981) Amino acid synthesis in photosynthesizing spinach cells. Effects of ammonia on pools sizes and rates of labelling from $^{14}CO_2$. *Plant Physiology*, **68**, 292-299.

Merola, A., Castaldo, R., De Luca, P., Gambardella, R., Musacchio, A., & Taddei, R. (1981) Revision of *Cyanidium caldarium*. Three species of acidophilic algae. *Giornale Botanico Italiano*, **115**, 189-195.

Miyachi, S. & Miyachi, S. (1985) Ammonia induces starch degradation in *Chlorella* cells. *Plant, and Cell Physiology*, **26**, 245-252.

Ohmori, M., Miyachi, S., Okabe, K. & Miyachi, S. (1984) Effects of ammonia on respiration, adenylate levels, amino acid synthesis and CO_2 fixation in cells of *Chlorella vulgaris* 11h in darkness. *Plant and Cell Physiology*, **25**, 749-756.

Rigano, C. & Conforti, T. (1967) Ecologia e distribuzione dell'alga unicellulare *Cyanidium caldarium* (Tilden) Geitler nei Campi Flegrei (Napoli). *Delpinoa*, **8-9**, 3-11.

Rigano, C. (1971) Studies on nitrate reductase from *Cyanidium caldarium*. *Archives f. Mikrobiologie*, **76**, 265-276.

Rigano, C., Aliotta, G., & Violante, V. (1974) Presence of high levels of nitrate reductase activity in *Cyanidium caldarium* grown on glutamate as the sole nitrogen source. *Plant Science Letters*, **2**: 277-281.

Rigano, C., Di Martino Rigano, V., Vona, V. & Fuggi, A. (1981) Nitrate reductase and glutamine synthetase activities, nitrate and ammonia assimilation, in the unicellular alga *Cyanidium caldarium*. *Archives of Microbiology*, **129**,110-114.

Romero, J.M, Coronil, T., Lara, C. & Guerrero, M.G. (1987) Modulation of nitrate uptake in *Anacystis nidulans* by the balance between NH_4^+ assimilation and CO_2 fixation. *Archives of Biochemistry and Biophysics*, **256**, 578-584.

Syrett, P.J. (1956) The assimilation of ammonia and nitrate by nitrogen starved cells of *Chlorella vulgaris*. IV. The dark fixation of carbon dioxide. *Physiologia Plantarum*, **9**, 165-171.

Thacker, A. & Syrett, P.J. (1972) The assimilation of nitrate and NH_4^+ by *Chlamydomonas reinhardii*. *New Phytologist*, **71**, 423-433.

Turpin, D.H. (1991) Effects of inorganic N availability on algal photosynthesis and carbon metabolism. *Journal of Phycology*, **27**, 14-20.

Can rising CO_2 alleviate oxidative risk for the plant cell?
Testing the hypothesis under natural CO_2 enrichment

M. BADIANI[1], A.R. PAOLACCI[1], A. D'ANNIBALE[1],
F. MIGLIETTA[2] AND A. RASCHI[2]

[1]*Dipartimento di Agrobiologia e Agrochimica, Universita' di Viterbo,*
Via S.C. De Lellis, I-01100 Viterbo, Italy.
[2]*I.A.T.A., Consiglio Nazionale delle Ricerche,*
Via G. Caproni,
8, I-50145 Firenze, Italy.

SUMMARY

Rising atmospheric CO_2 levels, by increasing the pCO_2/pO_2 ratio at the sites of photoreduction, could reduce the basal rate of O_2 activation and oxyradicals formation in plant tissues, thus alleviating the risk of oxidative stress of abiotic origin. Natural CO_2 springs of geothermal origin offer the opportunity of testing this hypothesis on both crop species and natural plant communities under otherwise undisturbed field conditions. In the present work, the contents of the major antioxidant enzymes and metabolites, indexes of oxidative damage to polyunsaturated fatty acids, and the level of one of the most aggressive reactive oxygen species, namely hydroxyl radical, were measured in foliar whole extracts obtained from wheat (*Triticum aestivum* L. cv. Mercia) plants grown in the natural high-CO_2 environment provided by a geothermal site in central Italy. The results obtained, also examined in the light of previous work on soybean (*Glycine max* Merrill cv. Cresir) plants grown under comparable high-CO_2 conditions (Badiani *et al.*, 1993), partly support the view that high CO_2 could reduce the risk of chloroplastic oxygen toxicity. However, it appears that the alleviation of the oxidative risk could vary not only with the plant species considered but could also depend on the CO_2 enrichment regime the plants are exposed to; species-specific detrimental effects on the prooxidant/antioxidant equilibrium could stem from the progressive suppression of energy-dissipating processes, such as photorespiration, under conditions of increasing CO_2. The data obtained suggest that natcral CO_2 enrichment affected also the extrachloroplastic antioxidant status in the plant material used. Among the antioxidant systems tested, the glutathione pool redox metabolism appeared to be the most heavily impacted by increasing CO_2 atmospheric concentration under field conditions.

Abbreviations used: AsA, ascorbic acid; APX, ascorbate peroxidase; C_a, atmospheric CO_2 concentration; CAT, catalase; DHAsA, dehydroascorbic acid; DHAR, dehydroascorbate reductase; DR, deoxyribose; DTNB, 5,5'-dithiobis; DW, dry weight; GAO, glycolic acid oxidase; GSH, glutathione; GSSG, glutathione disulfide; GR, GSSG reductase (NADPH); GPX, guaiacol peroxidase; i.r., infra-red; MDA, malondialdehyde; MDHAR, monodehydroascorbate reductase (NADH); PAGE, polyacrylamide gel electrophoresis; ROS, reactive oxygen species; RuBisCO, ribulose bisphosphate carboxylase/oxygenase; SOD, superoxide dismutase; TBA, thiobarbituric acid; TBARS, thiobarbituric acid reactive substances.

INTRODUCTION

The exploitation of carbon dioxide natural sources, such as geothermal sites (Miglietta & Raschi, 1993; Miglietta et al., 1993a, 1993b), provides an unequalled opportunity to study both short- and long-term adaptation mechanisms and pathways in plants, as well as to evaluate inter- and intraspecific diversity in responsiveness to increasing atmospheric CO_2 concentration (C_a). This could help our predictive ability and would allow the identification of adaptive proneness useful for the selection of agricultural species/varieties to be successfully employed in a changing CO_2 environment. Last, but not least, plant research in naturally CO_2-enriched environments could also improve our theoretical understanding of how environmental constraints, such as biotic and abiotic stressing factors, limit plant performance under current C_a.

Within the framework of a multidisciplinary investigation of the biology of plants exposed to carbon dioxide enrichment under the natural conditions occurring at CO_2 springs, we concentrated on the process of oxygen activation, which is a very frequent event during both normal and stress-altered plant metabolism (Halliwell & Gutteridge, 1989; Asada, 1994; Foyer, Descourvières & Kunert, 1994).

The roles of reactive oxygen species (ROS) in both normal and perturbed metabolism of humans, animals, plants and microorganisms have attracted considerable interest in recent years, even if little is known about their actual *in vivo* production rates within biological systems. In addition, once ROS are found to be associated with a particular pathogenic event, it is often difficult to assess whether they act as promoters/mediators of tissue injury and disease and/or if they have to be included among the effects of a given metabolic alteration (Halliwell & Gutteridge, 1989; Kehrer, 1993).

Despite the above uncertainties, cellular systems and organelles at risk from the standpoint of ROS-induced toxicity have been identified; amongst these, the chloroplast of higher plants. Even in the absence of promotive environmental conditions, this organelle, and in particular the membrane systems which support the photosynthetic energy conversion, are particularly exposed to the toxic effects of ROS. Given the intrinsically high probability of active oxygen species, it is not surprising that the highest levels of antioxidant metabolites (ascorbic acid, glutathione, carotenoids, tocopherol), interactive cooperative scavenging systems (the ascorbate-glutathione cycle) and effective repair enzymes (methionine sulfoxide reductase, thioredoxin) are all compartmentalized, even if not exclusively, within the chloroplast (Asada & Takahashi, 1987; Halliwell & Gutteridge, 1989; Foyer *et al.*, 1994; Foyer & Harbinson, 1994). It has been suggested that normally the levels of chloroplastic detoxifying enzymes, as well as their substrate concentrations and affinity, are sufficient to remove superoxide radical anion and H_2O_2 very efficiently, even under conditions of severe oxidative stress (Foyer *et al.*, 1991).

Among the well known consequences of CO_2 limitation of photosynthetic activity, two are of special importance for the promotion of oxidative stress in plant tissues; firstly, every circumstance able to reduce foliar CO_2 uptake further decreases the already low pCO_2/pO_2 ratio in the chloroplast, possibly leading to the accumulation of photoreducing power. This can cause an overflow of photosynthetic electrons, prone to partially reduce molecular oxygen. As a consequence, ROS can be formed which have been hypothesized as agents of photodamage and pigment photobleaching (Asada & Takahashi, 1987; Halliwell & Gutteridge, 1989;

Foyer & Harbinson, 1994; Polle & Rennenberg, 1994). Despite its proposed role in preventing the photoinactivation of electron transport in leaves, photorespiration is another intrinsic consequence of the low chloroplastic pCO_2/pO_2 prone to increase the cellular level of partially reduced O_2 forms, such as hydrogen peroxide (Wu, Neimanis & Heber, 1991). It is conceivable that oxidative stress occurring within the chloroplast could impact the whole plant performance (Alscher & Amthor, 1988).

Increasing C_a should increase the pCO_2/pO_2 ratio at the sites of photoreduction and this could reduce the basal rate of O_2 activation and oxyradicals formation. Also, it has been estimated that C_a of about 1500 $\mu L\ L^{-1}$, causing the carboxylating reaction of RuBisCO to outcompete its oxygenase activity, would virtually eliminate photorespiration in C_3 plants (Bowes, 1991). These effects of increasing C_a would be likely to result in a decreased cellular requirement of antioxidant activities.

The simple above-cited hypothesis is currently being tested on plant material growing at geothermal sites in central Italy and exposed to carbon dioxide concentrations ranging from about 350 to about 2400 $\mu L\ L^{-1}$ (Badiani et al., 1993).

In the present work, the contents of the major antioxidant enzymes and metabolites, indexes of oxidative damage to polyunsaturated fatty acids, and the level of one of the most aggressive ROS, hydroxyl radical, were measured in foliar whole extracts obtained from wheat (*Triticum aestivum* L. cv. Mercia) plants grown under increasing C_a regimes. The results obtained are also examined in the light of previous work (Badiani et al., 1993) done on a dicot species, soybean (*Glycine max* Merrill cv. Cresir), exposed to the natural high CO_2 environment provided by the geothermal sites.

MATERIALS AND METHODS

The carbon dioxide springs where the plant material was grown is located near (1 km) the town of Rapolano Terme (4000 residents, 250 m a.s.l., Siena district, central Italy); there is a small thermal bath, operating during the summer months, in the immediate surroundings of the experimental site. The site is distant from significant anthropogenic sources of gaseous air pollutants, such as large conurbations, motorways and industries; in this respect, it can therefore be regarded as a remote site.

On an annual basis, the climate at the site is considered to be suitable for the cultivation of winter wheat.

Previous measurements in the Rapolano spring, by means of a portable i.r. CO_2 analyser (EGM-1, PP Systems, Hitchin, UK), allowed the selection of two subsites whose average daytime C_a were 600 ± 21 (subsite 2xC) and 1000 ± 57 (subsite 3xC) $\mu L \ L^{-1}$. Both subsites were within about 30 m from the gas vents. A control subsite exposed to current C_a was also selected (subsite C), being located at about 1 km from the gas vents area (Miglietta *et al.*, 1993a, 1993b).

Measurements of $[H_2S]$ at the experimental site were made by H_2S-specific Dräger tubes (Dräger Aktiengesellschaft, Germany) (Miglietta & Raschi, 1993).

Plant material and growth conditions

Wheat seeds (*Triticum aestivum* L. cv. Mercia) were sown in rows 10 cm apart in 60 L containers filled with a loamy soil on 30 October 1992. A crop density of about 270 plants m^{-2} was obtained after seedling emergence.

The plants were grown according to standard agronomic practices. Water was automatically provided during the night. Carbon dioxide regimes were measured at plant height during most of the growing season in all the subsites by placing a sampling probe of the CO_2 analyser close to each of the containers. An automatic scanning device was built in order to switch air sampling between the CO_2 exposure variants every 15 min and CO_2 concentration data were recorded by a portable data logger (Delta-T Devices, Burwell, Cambridge, UK) by averaging over 5 min the readings made every minute. Meteorological data were also collected at each subsite throughout the plant growing season.

Leaf samples collection

Four samples, each consisting of 4-6 apparently healthy, fully expanded young flag leaves (about 1.5 g tissue each), excised from individual 180-d-old plants, were collected in the late morning for each CO_2 exposure variant. After excision, the leaves were immediately stored under dry ice and then, within 20 min, frozen at -80 °C until used. The frozen plant material was analysed within 24-48 hours after collection.

Analytical procedures

Previous control experiments showed that storage of wheat leaves at -80 °C had little or no effect on the contents/activities of antioxidant metabolites and enzymes. The stability of total glutathione at ultra-low temperatures was tested both in whole foliar tissues and in the acidic, protein-free supernatants obtained from them according to the procedure described below. Comparable slight depletions (5-10%) of total glutathione level were observed in both cases even after long-term (1 month) storage at -80 °C.

For antioxidant analysis, each sampled leaf, while still frozen, was divided longitudinally into 2 equal parts which were submitted to separate extraction. For the analysis of total glutathione ([GSH] + 2x[GSSG]), the catalytic assay proposed by Griffith (1985) was used, in view of its claimed "simplicity, excellent sensitivity and well understood specificity" (Griffith, 1985); briefly, the leaves were ground with a pestle in an ice-cold mortar in the presence of 80 mM trichloroacetic acid. The GSH of the resulting deproteinized extracts was then converted to GSSG by reacting with 5,5'-dithiobis (2-nitrobenzoic acid) (DTNB). The resulting GSSG was then reduced enzymically by NADPH and GSSG reductase (NADPH) (GR, EC 1.6.4.2; Sigma Chemical Co., Type IV, from bakers yeast, G4759) to regenerate GSH, which reacts again with DTNB.

The rate of formation of thionitrobenzoate dianion was followed spectrophotometrically at 412 nm and the assay was calibrated using GSH standards. If the deproteinized leaf extract was previously incubated with

2-vinylpyridine, GSH was derivatized and only GSSG was detected during the subsequent assay. Thus, total glutathione and GSSG concentrations could be separately determined and GSH concentration was calculated by the difference.

Aqueous extracts of leaves were obtained by grinding 4 g of tissue with a pestle in an ice-cold mortar with 16 mL of 50 mM sodium phosphate buffer (pH 7.00) containing 5mM 2-mercaptoethanol, 2mM dithiothreitol, 2mM Na_2-EDTA, 0.1 mM phenylmethylsulfonyl fluoride, 1% (w/v) bovine serum albumin, and 1% (w/v) polyvinylpyrrolidone (Polyclar AT^{TM}, Serva, Heidelberg, Germany). The supernatant obtained after filtration by four layers of cheesecloth and centrifugation at 39000 *g* for 30 min was used as an enzymic source (Badiani, De Biasi & Felici, 1990). On each of the above extracts the following measurements were performed: ascorbic (AsA) and dehydroascorbic (DHAsA) acids (Beutler, 1983); total ascorbate peroxidase activity (APX, EC 1.11.1.7; Asada, 1984); guaiacol peroxidase (GPX, EC 1.11.1.7; Badiani *et al.*, 1990); total superoxide dismutase activity (SOD, EC 1.15.1.1; Elstner, Youngman & Osswald, 1983); catalase (CAT, EC 1.11.1.6; Badiani *et al.*, 1990); soluble protein content (Protein Assay Reagent^{TM}, Pierce, Rockford, IL); GR (Smith, Vierheller & Thorne, 1988); DHAR [systematic name: GSH dehydrogenase (ascorbate), EC 1.8.5.1; Asada 1984]; monodehydroascorbate reductase (MDHAR; syn. ascorbate free radical reductase, EC 1.6.5.4; Borraccino, Dipierro & Arrigoni, 1989). Glycolic acid oxidase (GAO, EC 1.1.3.1) activity was measured spectrophotometrically by following the formation of glyoxylic acid phenylhydrazone in the presence of phenyl hydrazine hydrochloride and potassium glycolate (=1.7×10^4 at 324 nm) (Luna *et al.*, 1985).

The units of the enzymic activities were defined and calculated according to the respective assay methods. The crude extracts obtained from all the leaf samples were also submitted to anionic slab PAGE at pH 8.3, performed under constant current (20 mA) (Badiani *et al.*, 1990); samples containing from 300 to 600 μg soluble protein were loaded in each slab well. At the end of the run, GuaPod, CAT and SOD zymograms were stained on the slab gels as described by Badiani *et al.* (1990). Discrimination among Cu, Zn-, Mn- and Fe-containing SOD isoforms was performed by using CN^- or H_2O_2 as selective isozymes inhibitors as reported by Corpas *et al.* (1991).

The measurements of photosynthetic pigments were performed according to Lichtenthaler, Bach & Wellburn (1982) on *N,N*-dimethylformamide leaf extracts obtained following the methods described by Moran (1982) and Moran & Porath (1982).

The analysis of thiobarbituric-acid reactive substances (TBARS) was performed according to the procedure of Dhindsa & Matowe (1981) except that thiobarbituric acid (TBA) was dissolved in water prior to mixing with deproteinized plant extracts (Esterbauer & Cheeseman, 1990). The results were expressed in terms of malondihaldehyde (MDA) equivalents (=1.55×10^5 at 532 nm).

Hydroxyl radicals production was evaluated indirectly by measuring the extent of deoxyribose (DR) degradation (Halliwell, Grootveld & Gutteridge, 1988) after a 30 min incubation period at 37 °C in the presence of aliquots of plant extracts varying from 10 to 500 µL. This assay is based on the chromogenic properties of the molecular adduct resulting from the reaction of TBA with MDA, which is the main product of ˙OH-driven DR degradation. Appropriate blanks were run in order to discriminate between the MDA formed during DR degradation and that deriving from membrane peroxidative breakdown (see above).

Each of the above analytes was measured at least twice in the extracts obtained from each leaf sample. All the above procedures were performed at 4 °C. All reagents used were of analytical grade and were obtained from Aldrich (Steinheim, Germany), BDH (Poole, UK), Merck (Darmstadt, Germany), Serva (Heidelberg, Germany) and Sigma (St. Louis, USA).

Dry weight (DW) was determined by drying aliquots of 1 g leaf tissue in an oven at 80 °C until constant weight was reached.

RESULTS AND DISCUSSION

Meteorological data collected during the experimental periods at the three subsites, C, 2xC and 3xC, confirmed that the only environmental difference among them was represented by the atmospheric carbon dioxide partial pressure the plant material was exposed to (F. Miglietta and A. Raschi, unpublished observations). However, during day-time the surfaces of wheat leaves exposed to the 2xC and 3xC treatments were found to be somewhat warmer (1-2 °C) than control ones, and this was probably due to a higher degree of CO_2-dependent stomatal closure allowing a decreased heat dissipation capacity (F. Miglietta and A, Raschi, unpublished observations).

At the time of leaf samples collection, which corresponded to the stage of incipient ear formation, the plants submitted to natural CO_2 enrichment did not present any visible symptom of damage. However, they differed from those grown under normal C_a in terms of both onto-morphogenesis and phytomass accumulation (F. Miglietta and A. Raschi, unpublished observations). In comparison to the control growing under normal C_a, the wheat plants of subsites 2xC and 3xC showed a 30% increase in leaf area, 25-50% increase in leaf mass, an 80% increase in aboveground biomass and a slight reduction (10-20%) of foliar dry weight. At harvest, plants grown under high CO_2 showed a 60% increase in grain yield and ear mass. An acceleration in the onset of foliar senescence was also noticed in plants submitted to natural CO_2 enrichment (F. Miglietta and A. Raschi, unpublished observations).

A problem which may be encountered in the experimental use of carbon dioxide springs is the frequent presence of small amounts of non-condensable gaseous impurities in the CO_2 stream emitted from the vents (Axtmann, 1975). Among these contaminants, hydrogen sulfide is probably the most important. This is both because its concentration ratio to CO_2 in the geothermal fluids can be as high as 1:15 (Shinn *et al.*, 1976) and because it is known that H_2S can influence the metabolism and growth of sensitive plant species even at very low concentrations, especially under conditions of chronic exposure (Shinn *et al.*, 1976; Buwalda, De Kok & Stulen, 1993).

In the previous work on soybean (Badiani *et al.*, 1993), plants growing near the natural CO_2 springs of Grotte S. Stefano, Viterbo district, central Italy, were found to be exposed to substantial amounts of H_2S gushing out from the gas vents. This led to evaluating the extent of the possible H_2S

accumulation and toxicity in the experimental plant material (Rennenberg, 1984; Maas *et al.*, 1987; De Kok, Buwalda & Bosma, 1988; Maas & De Kok, 1988).

On the other hand, more favourable air quality conditions were encountered in Rapolano, where the subsites chosen for plant exposure to high CO$_2$ were found to be substantially free of H$_2$S contamination (data not shown, F. Miglietta and A. Raschi, unpublished observations). As a consequence, the analysis of reduced sulfur accumulation and toxicity was omitted in the wheat leaves used for the present experiment.

The wheat leaves exposed to the 3xC regime showed a reduction in the size of the glutathione pool, although it was not statistically significant (Fig. 1). However, discrimination between the reduced and oxidized form of this antioxidant revealed that CO$_2$ enrichment caused a dramatic decrease in the GSH level, paralleled by a concomitant increase of GSSG at 2xC and 3xC. Accordingly, a 10-fold increase in the GSSG/GSH ratio was observed in leaves exposed to CO$_2$ enrichment (Fig. 1). Interestingly, the leaves collected from subsite 2xC showed the highest GSSG/GSH ratio (see below).

On the whole, the observed effects of increasing CO$_2$ regimes on the glutathione pool and balance of wheat leaves were very similar to those previously obtained by exposing soybean leaves to natural CO$_2$ enrichment (Badiani *et al.*, 1993). It is remarkable that comparable CO$_2$ exposure regimes induced similar responses in species that are not only taxonomically unrelated but were also grown at different sites during different periods. Furthermore, since H$_2$S contamination was virtually absent at the Rapolano site, the similarity of the results obtained for the two species indirectly confirms that the presence of this gas in the geothermal emissions of the Grotte site did not influence the GSH metabolism of the soybean plants grown there (Badiani *et al.*, 1993). It should be mentioned that it is unknown whether the GSH measured in the leaves of the two species was solely glutathione and/or one of its homologues, homoglutathione in the case of soybean and hydroxymethylglutathione in the case of wheat (Buwalda *et al.*, 1993).

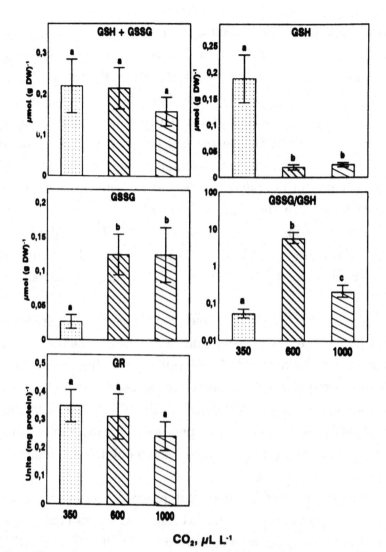

Fig. 1. The effects of increasing carbon dioxide on the foliar glutathione pool of wheat flag leaves. Acronyms denote: GSH, glutathione; GSSG, glutathione disulfide; GR, glutathione disulfide reductase. Each value is the mean of six replicates ± s.d. Values followed by the same letter are not significantly different at $P < 0.05$ %.

Despite the similar effects exerted by the treatment on the equilibrium between reduced and oxidized GSH forms, wheat foliar GR activity (Fig. 1) appeared to be not responsive to increasing CO_2, unlike the soybean foliar enzyme (Badiani *et al.*, 1993). In the latter species, in fact, the activity of this enzyme, regenerating GSH from GSSG at the expense of NADPH (Smith, Polle & Rennenberg, 1990), was strongly promoted by increasing C_a (Badiani *et al.*, 1993). By comparing the absolute titers measured in the two species grown under normal C_a, it was noted that, on a DW basis, the foliar GSH pool in wheat leaves was one order of magnitude lower than in soybean ones, whilst the GR specific activity was about 60 times higher (Fig. 1; Badiani et al., 1993). It is speculated that, unlike soybean, the lack of responsiveness of wheat GR to the CO_2-induced GSSG accumulation could be due to a more than adequate constitutive level of this regenerating enzyme. Despite this, the GR activity of high CO_2 leaves was unable to restore the GSSG/GSH value of control leaves. This failure could have been due to a lessened availability of photosynthetic reducing equivalents (Smith et al., 1990), caused in turn by an enhanced requirement of NADPH for carbohydrate synthesis at high C_a.

When applied to 2xC and 3xC leaves, the analytical procedure routinely employed to measure foliar ascorbate pool (AsA + DHAsA) failed to provide reliable values for DHAsA. As well as precluding the measurement of DHAsA, this problem did not allow the calculation of both the sizes of the ascorbate pool and the AsA/DHAsA ratios. The failure of the adopted method probably depended on the presence of extractable substances interfering with the preliminary dithiothreitol-driven reduction of endogenous DHAsA to AsA.

According to the original analytical procedure (Beutler, 1983), an increased carbohydrate status in the 2xC and 3xC leaves could have been involved in generating the above interference (Stitt, 1991; F. Miglietta, personal communication).

In wheat leaves, the specific activities of the enzymes involved in the scavenging of $\cdot O_2^-$ and H_2O_2, namely SOD, APX, GPX and CAT, as well as DHAR, presented the common features of being increased by the 2xC treatment and of reverting to control values under the 3xC growth conditions (Fig. 2). For most enzymes, the extent of the 2xC induction was limited, and only in the case of APX a significant ($P < 0.05\%$) difference from control values was observed. As expected, and unlike soybean (Badiani et al., 1993), PAGE analysis at pH 8.3 did not reveal any effect of increasing CO_2 on the isozymic patterns of SOD, GPX and CAT (data not shown).

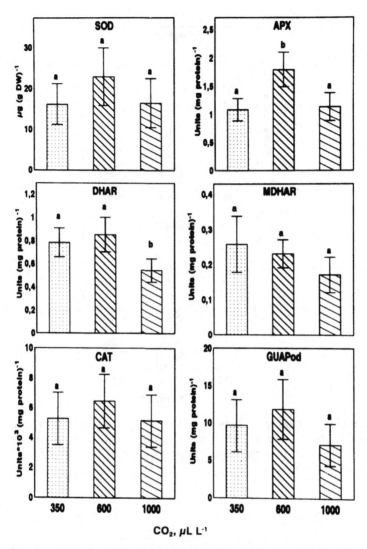

FIG. 2. Active oxygen-scavenging and antioxidant-regenerating enzymes in wheat flag leaves grown under increasing carbon dioxide partial pressure. Acronyms denote: APX, ascorbate peroxidase; MDHAR, monodehydroascorbate reductase; DHAR, dehydroascorbate reductase; SOD, superoxide dismutase; CAT, catalase; GPX, guaiacol peroxidase. Measurement replication and statistical significance levels as in Fig. 1.

The GAO extractable capacity measured in wheat leaves and in soybean leaflets coherently suggested that increasing CO_2 progressively depressed photorespiration (Fig. 3).

As increases in antioxidant enzyme activities are often regarded as compensatory responses (Halliwell & Gutteridge, 1989; Scandalios, 1990; Foyer *et al.*, 1994), on the whole the data of Fig. 2 indirectly suggest the

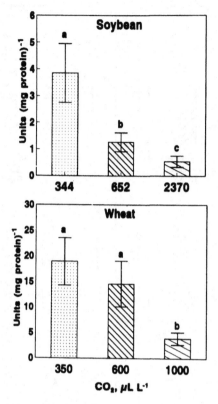

FIG. 3. Glycolic acid oxidase activity in leaves of dicot and monocot species grown under increasing carbon dioxide partial pressure near geothermal sites in central Italy. Each value is the mean of eight (soybean) or six (wheat) replicates ± s.d. Statistical significance levels as in Fig. 1.

intriguing possibility that the 2xC regime, but not the 3xC one, caused a perturbation in the normal balance between formation and removal of ROS in wheat leaves. This same suggestion originated from the results obtained for the GSSG/GSH ratio (Fig. 1), whose increase is often regarded as an indicator of oxidative stress (Lopez-Torres *et al.*, 1993).

In search of direct evidence for the above hypothesis, certain widely accepted cellular indicators of oxidative stress and damage, such as the extent of hydroxyl radical production (Halliwell *et al.*, 1988), the TBARS titer (Esterbauer & Cheeseman, 1990) and the contents of photosynthetic pigments (Young, 1991) were measured in wheat leaves submitted to CO_2 enrichment.

Indeed, the extent of the OH-driven DR degradation, thought to be directly proportional to the rate of endogenous hydroxyl radical formation, supported the view that plant tissues facing the 2xC treatment were particularly exposed to oxyradical attack (Fig. 4). The possible over production of ROS, such as OH, O_2 and H_2O_2, in foliar tissues exposed to 2xC appeared not to exceed the cellular homoeostatic capacity, since an index of membrane lipid breakdown, such as TBARS, clearly tended to decline as C_a increased (Fig. 4). Furthermore, no evidence of pigment degradation was obtained in wheat leaves exposed to CO_2 enrichment; on the contrary, the chlorophyll *a*, chlorophyll *b* and carotenoids foliar contents increased at 2xC and then declined to the control value in the 3xC leaves (Fig. 4).

CONCLUSIONS

The question in the title remains open. Moreover, it is not possible to draw general conclusions from the data obtained since only two plant species were examined and their leaves were collected and analysed only once during the plants' life cycles.

Caution should be also exerted in extrapolating the CO_2-enrichment effects observed on selected leaves to the whole plant.

However, certain suggestions for further work can be made considering the results obtained from wheat and soybean leaves exposed to natural CO_2 enrichment together (see also Badiani *et al.*, 1993):

a) part of the evidence obtained indirectly supports the view that, in a range of C_a largely encompassing also long-term scenarios, high CO_2 could reduce the risk of chloroplastic oxygen toxicity. This alleviation could result from both the stimulation of the photosynthetic electron withdrawal operated by the Calvin-Benson cycle and from a gradual suppression of photorespiration;

b) however, the above hypothesis appears to be too simple to account for all the experimental observations. To begin with, the alleviation of the oxidative risk could vary not only with the plant species considered but

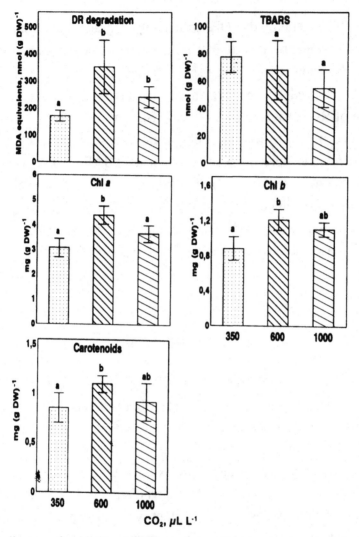

FIG. 4. Indicators of oxidative stress/damage in wheat leaves grown under increasing CO$_2$ partial pressure. Acronyms denote: DR, 2-deoxyribose, Chl, chlorophyll; MDA, malondialdehyde; TBARS, thiobarbituric acid reactive substances. Measurement replication and statistical significance levels as in Fig. 1.

could also depend, in a complex fashion, on the CO$_2$ enrichment regime the plants are exposed to.

It has been suggested that detrimental effects on photosynthetic activity could occur beyond a certain C$_a$ value (Bowes, 1991; Stitt, 1991). This is due to the intervention of limiting factors, among which are the activity of RuBisCO, the cellular regeneration capacity of its CO$_2$-O$_2$ acceptor, the end-products inhibition and mineral nutrients availability.

In soybean, the C_a threshold for the onset of these limitations would be particularly high, since this species, under conditions of high CO_2, seems able to avoid starch accumulation by increasing leaf thickness, dry weight, number of mesophyll cells and size and number of chloroplasts, thus reducing the need to down-regulate RuBisCO activity (Bowes, 1991).

Other species, however, such as *T. aestivum*, could lack this array of compensating mechanisms; as a result, once a critical CO_2 threshold is exceeded, NADPH accumulation and the associated risk of oxygen toxicity could rise again, as under normal C_a. Thus, it would be of interest to evaluate these short-term adaptive aspects in different species exposed to CO_2 enrichment under field conditions;

c) species-specific detrimental effects on the prooxidant/antioxidant equilibrium could well stem from the progressive suppression of an energy dissipating process such as photorespiration under conditions of increasing CO_2. In this respect, it could be worth comparing the behaviour of antioxidant systems in C3, C4 and intermediate species;

d) the data obtained suggest that CO_2 enrichment also affected the extrachloroplastic antioxidant status. Hence, studies on purified subcellular components are needed, with particular reference to the peroxisomal, mitochondrial and apoplastic fractions;

e) among the antioxidant systems tested, the glutathione pool redox metabolism appeared to be the most heavily impacted by increasing CO_2 atmospheric concentration. On the one hand, this suggests the need to test additional exposure variants within realistic CO_2 concentration ranges (e.g. 350-700 µL L^{-1}) in order to obtain more accurate dose-response curves and, possibly, critical CO_2 levels. On the other hand, further investigation is required aiming at clarifying the possible reason(s) for the high degree of responsiveness of this major antioxidant system to enhanced CO_2.

ACKNOWLEDGEMENTS

M.B. gratefully acknowledges Dr L.J. De Kok and his colleagues at the University of Groningen for the very useful suggestions about H_2S metabolism; many thanks to Dr. A. Giuntoli and Dr. B. Rapi for providing biometric data and to Prof G. Giovannozzi Sermanni for his encouragement during this work. The cost of the research has been partially covered by EU, Programme Environment, contract EV5V-CT93-0432.

REFERENCES

Alscher, R.G. & Amthor, J.S. (1988) The physiology of free-radical scavenging: maintenance and repair processes. In *Air Pollution and Plant Metabolism* (eds. S. Schulte-Hostede, N.M Darrall, L.W. Blank & A.R. Wellburn), pp. 94-115. Elsevier Applied Science, London and New York.

Asada, K. (1984) Chloroplasts: formation of active oxygen and its scavenging. *Methods in Enzymology*, 105, 422-429.

Asada, K. (1994) Production and action of active oxygen species in photosynthetic tissues. In *Causes of Photooxidative Stress and Amelioration of Defense Systems in Plants* (eds. C.H. Foyer & P.M. Mullineaux), pp. 77-104. CRC Press, Boca Raton.

Asada, K. & Takahashi, M. (1987) Production and scavenging of active oxygen in photosynthesis. In *Photoinhibition* (eds. D.J. Kyle, C.B. Osmond & C.J. Arntzen), pp. 227-287. Elsevier, Amsterdam.

Axtmann, R.C. (1975) Emission control of gas effluents from geothermal power plants. *Environmental Letters*, 8, 135-146.

Badiani, M., De Biasi, M.G. & Felici, M. (1990) Soluble peroxidase from winter wheat seedlings with phenoloxidase-like activity. *Plant Physiology*, 92, 489-494.

Badiani, M., D'Annibale, A., Paolacci, A.R., Miglietta, F. & Raschi, A. (1993) The antioxidants status of soybean (*Glycine max* Merrill) leaves grown under natural CO_2 enrichment in the field. *Australian Journal of Plant Physiology*, 20, 275-284.

Beutler, H.-O. (1983) L-ascorbate and L-dehydroascorbate. In *Methods of Enzymatic Analysis* (ed. H.U. Bergmeyer.), 2nd Ed. Vol. 8., pp. 376-385. Verlag Chemie, Weinheim.

Borraccino, G., Dipierro, S. & Arrigoni, O. (1989) Interaction of ascorbate free radical reductase with sulfhydryl reagents. *Phytochemistry*, 28, 715-717.

Bowes, G. (1991) Growth at elevated CO_2: photosynthetic responses mediated through rubisco. *Plant, Cell and Environment*, 14, 795-806.

Buwalda, F., De Kok, L.J. & Stulen, I. (1993) Effects of atmospheric H_2S on thiol composition of crop plants. *Journal of Plant Physiology*, 142, 281-285.

Corpas, F.J., Sandalio L.M., Palma J.M., Leidi E.O., Hernandez, J.A., Sevilla, F. & del Rio, L.A. (1991) Subcellular distribution of superoxide dismutase in leaves of ureide-producing leguminous plants. *Physiologia Plantarum*, 82, 285-291.

De Kok, L.J., Buwalda, F. & Bosma, W. (1988) Determination of cysteine and its accumulation in spinach leaf tissue upon exposure to excess sulfur. *Journal of Plant Physiology,* **133,** 502-505.

Dhindsa, R.S. & Matowe, W. (1981) Drought tolerance in two mosses: correlated with enzymatic defence against lipid peroxidation. *Journal of Experimental Botany,* **32,** 79-91.

Elstner, E.F., Youngman R.J. & Osswald, W. (1983) Superoxide dismutase. In *Methods of Enzymatic Analysis* (ed. H.U. Bergmeyer.), 2nd Ed., Vol. 3, pp. 293-302. Verlag Chemie, Weinheim.

Esterbauer, H. & Cheeseman, K.H. (1990) Determination of aldehydic lipid peroxidation products: malonaldehyde and 4-hydroxynonenal. *Methods in Enzymology,* **186,** 407-421.

Foyer C.H., Descourvières P. & Kunert K.J. (1994) Protection against oxygen radicals: an important defense mechanism studied in transgenic plants. *Plant, Cell and Environment,* **17,** 507-523.

Foyer C.H. & Harbinson, J. (1994) Oxygen metabolism and the regulation of photosynthetic electron transport. In *Causes of Photooxidative Stress and Amelioration of Defense Systems in Plants* (eds. C.H. Foyer & P.M. Mullineaux), pp. 1-42. CRC Press, Boca Raton.

Foyer, C., Lelandais, M., Galap, C. & Kunert, K.J. (1991) Effects of elevated cytosolic glutathione reductase activity on the cellular glutathione pool and photosynthesis in leaves under normal and stress conditions. *Plant Physiology,* **97,** 863-872.

Griffith, O.W. (1985) Glutathione and glutathione disulphide. In *Methods of Enzymatic Analysis* (ed. H.U. Bergmeyer.), 2nd Ed., Vol. 8., pp. 521-529. Verlag Chemie, Weinheim.

Halliwell, B., Grootveld, M. & Gutteridge, J.M.C. (1988) Methods for the measurement of hydroxyl radicals in biochemical systems: deoxyribose degradation and aromatic hydroxylation. *Methods of Biochemical Analysis,* **33,** 59-90.

Halliwell, B. & Gutteridge, J.M.C. (1989) *Free Radicals in Medicine and Biology,* 2nd Edn, pp. 277-289. Clarendon Press, Oxford.

Kehrer, J.P. (1993) Free radicals as mediators of tissue injury and disease. *Critical Reviews in Toxicology,* **23,** 21-48.

Lichtenthaler, H.K., Bach, T.J. & Wellburn, A.R. (1982) Cytoplasmic and plastidic isoprenoids compounds of oat seedlings and their distinct labelling from ^{14}C-Mevalonate. In *Biochemistry and Metabolism of Plant Lipids* (eds. J.F.G.M. Wintermans & P.J.C. Kuiper), pp. 489-500. Elsevier Biomedical Press, Amsterdam.

Lopez-Torres, M., Perez-Campo, R., Cadenas, S., Roias, C. & Baria, G. (1993) A comparative study of free radicals in vertebrate - II. Non-enzymatic antioxidants and oxidative stress. *Comparative Biochemistry and Physiology,* **105B,** 757-763.

Luna, M., Badiani, M., Felici, M., Artemi, F. & Giovannozzi Sermanni, G. (1985) Selective enzyme inactivation under water stress in maize (*Zea mays L.*) and wheat (*Triticum aestivum* L.) seedlings. *Environmental and Experimental Botany*, **25**, 153-156.

Maas, F.M., De Kok L.J., Peters J.L. & Kuiper, P.J.C. (1987) A comparative study on the effects of H_2S and SO_2 fumigation on the growth and accumulation of sulphate and sulphydryl compounds in *Trifolium pratense* L., *Glycine max* Merr. and *Phaseolus vulgaris* L. *Journal of Experimental Botany*, **38**, 1459-1469.

Maas, F.M. & De Kok, L.J. (1988) *In vitro* NADH oxidation as an early indicator for growth reduction in spinach exposed to H_2S in the ambient air. *Plant and Cell Physiology*, **29**, 523-526.

Miglietta, F. & Raschi, A. (1993) Studying the effects of elevated CO_2 in the open in a naturally enriched environment in Central Italy. *Vegetatio*, **104/105**, 391-400.

Miglietta, F., Raschi, A., Bettarini, I., Resti, R. & Selvi, F. (1993a) Natural CO_2 springs in Italy: A resource for examining long-term response of vegetation to rising atmospheric CO_2 concentrations. *Plant, Cell and Environment*, **16**, 873-878.

Miglietta, F., Raschi, A., Bettarini, I., Badiani, M. & van Gardingen, P. (1993b) Carbon dioxide springs and their use for experimentation. In *Design and Execution of Experiments on CO_2 Enrichment* (eds. E.D. Schulze & H.A. Mooney), Ecosystems Research Report No. 6, pp. 393-403. Commission of the European Communities, Brussels.

Moran, R. (1982) Formulae for the determination of chlorophyllous pigments extracted with *N,N*-dimethylformamide. *Plant Physiology*, **69**, 1376-1381.

Moran, R. & Porath, D. (1982) Chlorophyll determination in intact tissues using *N,N*-dimethylformamide. *Plant Physiology*, **65**, 478-479.

Polle, A. & Rennenberg, H. (1994) Photooxidative stress in trees. In *Causes of Photooxidative Stress and Amelioration of Defense Systems in Plants* (eds. C.H. Foyer & P.M. Mullineaux), pp. 199-218. CRC Press, Boca Raton.

Rennenberg, H. (1984) The fate of excess sulphur in higher plants. *Annual Review of Plant Physiology*, **35**, 121-153.

Scandalios, J.G. (1990) Response of plant antioxidant defense genes to environmental stress. In *Genomic Responses to Environmental Stress* (ed. J.G. Scandalios), pp. 1-41. Academic Press, San Diego.

Shinn, J.H., Clegg, B.R., Stuart, M.L. & Thompson, S.E. (1976) Exposure of field-grown lettuce to geothermal air pollution. Photosynthetic and stomatal responses. *Environmental Sciences and Health*, **A11**, 603-612.

Smith, I.K., Vierheller, T.L. & Thorne, C.A. (1988) Assay of glutathione reductase in crude tissue homogenates using 5,5'-dithiobis (2-nitrobenzoic acid). *Analytical Biochemistry*, **175**, 408-413.

Smith, I.K., Polle, A. & Rennenberg, H. (1990) Glutathione. In *Stress Responses in*

Plants: Adaptation and Acclimation Mechanisms (eds. R.G. Alscher & J.R. Cumming), pp. 201-215. Wiley-Liss, New York.

Stitt, M. (1991) Rising CO_2 levels and their potential significance for carbon flow in photosynthetic cells. *Plant, Cell and Environment*, **14**, 741-762.

Wu, J., Neimanis, S. & Heber, U. (1991) Photorespiration is more effective than the Mehler reaction in protecting the photosynthetic apparatus against photoinhibition. *Botanica Acta*, **104**, 283-291.

Young, A.J. (1991) The photoprotective role of carotenoids in higher plants. *Physiologia Plantarum*, **83**, 702-708.

Increasing concentrations of atmospheric CO_2 and decomposition processes in forest ecosystems

P. INESON[1] AND M. F. COTRUFO[2]

[1] *Merlewood Research Station, Grange-over-Sands, Cumbria LA11 6JU, UK.*
[2] *Division of Biological Sciences, Institute of Environmental and Biological Sciences, Lancaster University, Lancaster, LA1 4YQ, UK.*

SUMMARY

The impacts of elevated atmospheric CO_2 concentration on decomposition processes are reviewed, with consideration given to both the direct and indirect effects which CO_2 may exert on soil processes.

Present work indicates that elevated concentrations of CO_2 affect litter decomposition through changes in litter quality.

In particular, elevated CO_2 decreases litter N concentrations with a resulting increase in C/N and lignin/N ratios, leading to a slow down in litter decomposition rates.

Such observations are true for both leaf and root litters, and for laboratory as well as field incubations. CO_2 treatment does not appear to exert any effect on litter decay rates when there are no measurable changes in leaf chemical composition.

Furthermore, it has been identified that elevated CO_2 is unlikely to exert any direct effects on soil biological activity, and that only effects mediated *via* changes in plants are likely to occur.

Although there is currently debate about the effect which increased C inputs to soils resulting from elevated CO_2 may have on the turnover rates of existing soil organic matter, the overall conclusion is that soil C stores are increasing, and will continue to increase, under higher CO_2 concentrations.

INTRODUCTION

If accurate predictions of the influence of elevated CO_2 on terrestrial carbon (C) balance and on the global C cycle are to be made, then more information is needed about the development of above- and below-ground C stores under different atmospheric CO_2 concentrations. The extent to which terrestrial ecosystems are able to 'buffer' against rising atmospheric CO_2 concentrations by storing more or less C needs to be assessed, together with the changes which are currently occurring. Soil C stores are the resultant between net primary production and decomposition, and we consider here the known and potential effects of elevated CO_2 concentrations on both productivity and decomposition.

Human activity has overwhelmed the natural global CO_2 balance, inducing an increase in atmospheric CO_2 concentration from 280 ppm, of pre-industrial levels, to 354 ppm in 1990 (World Resources Institute, 1992). In recent C budget analyses, the fate of large fractions of anthropogenic C emissions remains unaccounted for, and this quantity is commonly referred to as the 'missing carbon' sink, having recently been estimated at 0.4-4.3 Pg C per year (Gifford, 1994). Attention has been directed towards terrestrial ecosystems and their role in balancing global C budgets, with the suggestion that C may be accumulating in soils in the temperate zone (Houghton, 1993). In particular, there is evidence of increasing global forest productivity due to CO_2 fertilization, and that forest ecosystems may be currently acting as a net sink for C (Tans, Fung & Takahashi, 1990). In forest ecosystems the main source of organic matter entering the soil is plant litter, with leaf litter accounting for 22 to 81% of total annual litter production (Mason, 1977), whilst perennial parts contribute from 12 to 53% of litter production (Rodin & Bazilevich, 1967). The fine root system is also of major importance in terms of forest productivity and turnover, and has been estimated to account for between 8 and 67% of primary production in forests (Santantonio & Grace, 1987). Decomposition of plant litter is a key process in nutrient cycling in forest ecosystems and, through this process, nutrients immobilized in the detritus are mineralized and released in forms suitable for plant uptake. In nutrient poor ecosystems, litter decomposition may become the controlling step for nutrient cycling and forest productivity (Flanagan & Van Cleve, 1983). Any factor influencing either the production or decomposition rates of litter in forest ecosystems will affect the total C stored, and we consider here the ways in which elevated CO_2 may influence these processes.

CHANGES IN LITTER QUANTITY

Increased plant productivity under elevated CO_2 concentrations has been observed for a wide range of plant species and growth conditions. In a major synthesis of over 60 controlled CO_2-exposure studies, Wullschleger, Post & King (1995) reported the growth response of a large number of forest tree species to elevated CO_2. These authors summarized the responses as a relative response ratio (RRR), and reported a mean RRR for the dataset of 1.32 (corresponding to an increase of 32%) in total plant biomass across the 398 observations analysed (Fig. 1). The variability in RRR was high, with different responses for different species, yet the growth enhancement under elevated CO_2 did not appear to be greatly constrained by the availability of water or nutrients. Low soil N supply only reduced the RRR from 1.36 to 1.28, when compared to adequate N supply; similarly, under water deficient conditions the RRR was 1.18, very close to the value of 1.22 found with adequate water supply. An important conclusion arising from this review was that the 32% growth increase observed under elevated CO_2 appeared equally allocated to all plant organs, with 33%, 29% and 38% increases being calculated respectively for leaf, stem + branch and total root dry mass.

The data included by Wullschleger et al. (1995) were derived from short-term controlled CO_2 enrichment studies, and extrapolation to the global scale must be viewed with caution. The growth responses of woody perennial species under long-term exposure to elevated atmospheric CO_2 still needs to be assessed, but the technical difficulties associated with fumigation severely limit experimentation with mature trees.

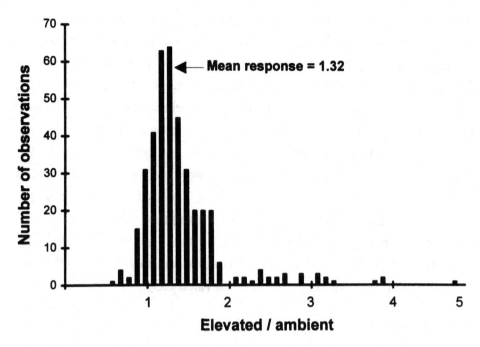

FIG. 1. Frequency distribution for the response of total plant dry mass (above- and below-ground) to an approximate doubling of atmospheric CO_2 concentration. Data from Wullschleger *et al.*, 1995.

Additionally, the assessment of long-term growth responses of woody species to increasing CO_2 really requires a quantitative assessment of nutrient cycling, and C and water balance in intact forest ecosystems (Mooney *et al.*, 1991). Recent examinations of tree ring chronologies have indicated increases in growth of mature trees which correlate with the rising atmospheric CO_2 concentrations across a wide range of forests (Luxmoore, Wullschleger & Hanson, 1993). However, the influence of climatic conditions and N deposition on forest growth may interact with increasing CO_2, and growth responses observed in most tree-ring records exeeded those expected from CO_2 enrichment alone (Mooney *et al.*,1991).

Elevated concentrations of atmospheric CO_2 lead to a more rapid growth of trees (Eamus & Jarvis, 1989), and may extend the growing season, resulting in delayed senescence. Studies on the effects of elevated CO_2 on ecosystem processes, conducted in Arctic tundra and salt marsh ecosystems, have both shown changes in plant phenology induced by increased CO_2, with slower rates of senescence (Curtis *et al.*, 1989a; Tissue & Oechel, 1987). In contrast, other studies on the effects of elevated CO_2 on plant phenology have observed early leaf senescence (Bhattacharya *et al.*, 1985; St. Omer & Horvath, 1983). Alterations in the timing of leaf senescence may have consequences for subsequent decomposition rates, which in turn may have an impact on nutrient cycling.

Even if nutrient limitations come into play, and act as a brake on the fertilizing effect of CO_2, the overall conclusion is that forest ecosystems will produce more litter under elevated CO_2, and that increases in soil C stores are anticipated. However, if the litter produced under elevated CO_2 has altered litter quality (Melillo, 1983), then this may modify this conclusion. It is therefore necessary to consider how litter quality and decomposition rates are modified under elevated CO_2.

CHANGES IN LITTER QUALITY

Dilution of N concentration of plant tissue associated with CO_2 enrichment treatment has been observed for trees (Brown, 1991; El-Kohen, Rouhier & Mousseau, 1992; Norby, Pastor & Melillo, 1986a; O'Neill, Luxmoore & Norby, 1987) and other types of vegetation (Barnes & Pfirrmann, 1992; Coleman, McConnaughay & Bazzaz, 1993; Johnson & Lincoln, 1991; Overdieck & Reining, 1986; Wong, 1979). Several explanations have been given for this effect, but increased plant growth, resulting from CO_2 fertilization under N deficient conditions, appears to cause an unbalanced uptake of N together with an increase in N-use efficiency (Norby *et al.*,1986a; O'Neill *et al.*,1987). Coleman *et al.* (1993) suggested that the CO_2-induced reduction in plant N concentration may be a size-dependent phenomenon resulting from accelerated plant growth. However, Ågren (1994) has argued that the effect of size is indirect, and pointed out that caution is needed in interpreting the effects due solely to elevated CO_2 from those arising from the interaction between CO_2 and nutrition. Additionally, CO_2 enrichment may induce changes in the distribution pattern of N within the plant, with greater proportions of N being translocated into metabolically active tissues (Norby *et al.*,1986a).

If these observed changes in N content of plants in response to elevated CO_2 have any significance in natural ecosystems, it should be possible to find evidence of N changes in existing vegetation "time series ". In a study of changes in N and S concentrations of herbarium specimen leaves, from AD 1750 to the present time, Peñuelas & Matamala (1990) reported a close relationship between leaf N content and atmospheric CO_2 concentrations, with a progressive decrease in leaf N content with time, paralleling recorded increases in CO_2 concentrations (Fig. 2). The changes in N concentrations differed between species, but a relative mean reduction from 144% in the year 1750 relative to 100% for the present day was observed. No similar significant time trends were observed for leaf S contents.

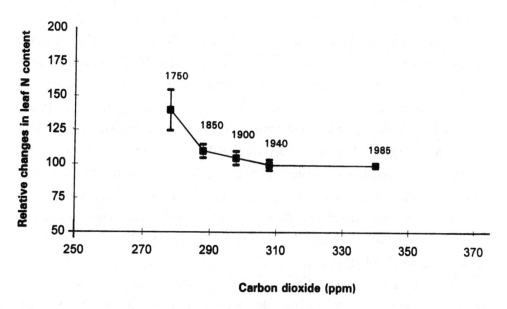

FIG. 2. Relative changes in N content of herbarium sample across 14 species of trees, shrubs and herbs. Data from Peñuelas & Matamala (1990).

Despite the overall conclusions of Wullschleger *et al.* (1995), plant responses to CO_2 enrichment have been shown to be influenced by the nutrient status of soil (Bazzaz & Miao, 1993). However, decreased N concentrations in plant material exposed to elevated CO_2 have been commonly reported at both low and high levels of soil nutrition, suggesting either changes in the physiology of the plants (El-Kohen *et al.*, 1992) or alterations in rhizosphere interactions (Diaz *et al.*, 1993).

The majority of the observations on changes in plant tissue quality induced by elevated CO_2 refer to green plant parts, and it has been suggested that senescent tissue may behave differently (O'Neill, 1994). The little data available on changes in N concentration of leaf litters show contrasting results, with no effects being reported for oak (Norby *et al.*,1986a) and estuarine marsh plant (Curtis, Drake & Whingham, 1989b) litters, whereas a significant decrease in N concentrations of leaf litter tissues exposed to enriched CO_2 atmosphere was observed for other tree species (Cotrufo, Ineson & Rowland, 1994; Coûteaux *et al.*, 1991).

Reports on C/N ratios of tissues derived from plants exposed to elevated CO_2 have been rare, with studies differring in experimental condition, concentration of CO_2, etc. (Table 1). However, 20 observations have been reported in the literature and the C/N ratios of plant tissues grown under elevated CO_2 were regressed against C/N ratios of the same plant tissue grown under ambient CO_2 regimes, and are presented in Fig. 3.

A 24% increase in the C/N ratio of plant tissue generated under elevated CO_2 was found, when compared to tissue derived from ambient CO_2 environments. For this analysis, the data from the 20 observations were pooled together regardless of species, N fertilization treatment and plant organ. Of the 13 species represented, 7 were woody; only 3 observations were under low N soil status, and a wide range of organs were represented: fine roots ($n=6$), leaves ($n=2$), leaf litter ($n=7$), shoot ($n=2$) and whole plant ($n=3$).

The effects of elevated CO_2 on nutrients other than N are somewhat more confused. There is evidence that P concentrations (unlike N) in plant tissue do not decline under elevated CO_2 levels (Oberbauer *et al.*, 1986; Ownesby, Coyne & Aven, 1993). A possible explanation for this difference in behaviour between N and P is that, under elevated CO_2, plant N-use efficiency increases and N uptake appears to remain constant, whereas plant requirements for P have been shown to increase (Conroy & Hocking, 1993). However, decreased P concentrations in plants grown under enriched CO_2 have been observed (Barnes & Pfirrmann, 1992; Woodin *et al.*, 1992). Tissue concentrations of K, Mg and Ca were found to be

TABLE 1. References used for the analysis of the effects of elevated CO_2 on plant tissues' C/N ratio

Reference	Species	[CO_2] ppm	Interactions	Organs
Barnes & Pfirrmann (1992)	*Raphanus sativus*	350/7500₃	exposure	Shoot;Root
Chu, Coleman & Mooney (1992)	*Raphanus sativus*	0/600	-	Shoot;Root
Cotrufo *et al.* (1994)	*Betula pubescens* *Acer pseudoplatanus* *Fraxinus excelsior*	350/600	-	Leaf litter
Cotrufo & Ineson (1995)	*Betula pendula* *Picea sitchensis*	350/600	N supply	Fine root
Coûteaux *et al.* (1991)	*Castanea sativa*	350/70	-	Leaf litter
Curtis *et al.* (1989b)	*Scirpus olneyi* *Spartina patens*	350/686*	-	Leaf; Litter
Norby *et al.* (1986a)	*Quercus alba*	362/690*	-	Leaf litter
Overdieck, Reid & Strain (1988)	*Vigna unguiculata* *Abelmoschus esculentus* *Raphanus sativus*	0/350/650	-	Different

* Values were rounded to the nearest 50 ppm for the analysis

unaffected in seedlings of the tree *Maranthes corymbosa* grown in fertile soil and exposed to 700 ppm CO_2 when compared to control (350 ppm CO_2) trees (Berryman, Eamus & Duff, 1993). Increased nutrient (P, K, Fe, Cu and Al) uptake in tree seedlings grown under elevated CO_2, which resulted in no changes in tissue nutrient concentrations, has been observed in nutrient poor soil (Norby, O'Neil & Luxmoore, 1986b).

In contrast, reduced K, Mg and Ca concentrations were observed for *Calluna vulgaris* (L.) plants raised in nutrient poor peat soil under elevated atmospheric CO_2 levels (Woodin *et al.*, 1992).

No significant changes in P, K, Ca or Mg concentrations were reported for tree leaf litters generated under elevated CO_2 atmosphere when compared to control litters (Cotrufo *et al.*,1994).

Increased concentrations of lignin and phenolics have been reported in plants exposed to high levels of atmospheric CO_2 (Cipollini, Drake & Whigham, 1993; Cotrufo *et al.*,1994; Melillo, 1983). However, it has been

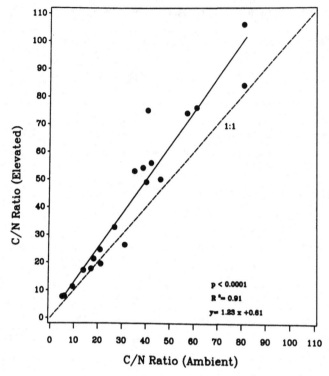

FIG. 3. C/N ratio of plant tissues grown at ambient and enriched CO$_2$ levels. Data from studies reported in Table 1.

suggested that such an accumulation of lignin, as well as of other phenolic compounds, may be due to a limitation of nutrient supply, rather than a direct effect of CO$_2$ enrichment (Lambers, 1993).

In contrast, Norby *et al.* (1986a) found a lower amount of lignin, and a decrease in lignin/N ratio, in seedlings of white oak grown in elevated CO$_2$, yet the values shown for both treatments were very low (below 5 % of ash-free dry weight for the lignin and 6 for the lignin/N ratio).

In a study on the secondary chemical composition of *Salix myrinifolia* (Salisb.), an interaction between CO$_2$ treatments, soil type and clonal origin was observed with respect to tissue phenolic concentrations, but it appeared that increased CO$_2$ lowered the content of phenolic glucosides, whilst increasing polymeric phenolic concentrations (Julkunen-Tiitto, Tahvanainen & Silvola, 1993).

Increased CO$_2$ enhanced the concentration of tannins in plant tissue, with the extent of response being species-specific; a 40% increase in tannin concentrations was reported for white oak exposed to double CO_2 levels (Norby et al.,1986a), whilst increased tannin concentrations of 22%, 87% and 90% were observed respectively for oak, sugar maple and aspen seedlings grown under 650 ppm CO_2, when compared to seedlings grown under ambient CO_2 regimes (Lindroth, Kinney & Platz, 1993). Enriched CO_2 has also been associated with an increase in leaf thickness, due to an increased number of palisade cells (Vu, Allen & Bowes, 1989).

From the observations reported above there is ample evidence that enhanced CO_2 levels may induce major changes in chemical composition and quality of plant tissues, but the direction and magnitude of the effects is highly species-specific and dependent on other environmental factors, such as soil nutrient status, and plant growth conditions.

EFFECTS ON DECOMPOSITION

Berg (1984), among others, has observed that initial N litter content is an important rate-regulating factor in the early stages of decomposition, whilst lignin concentration becomes more important in later stages. Similarly, Mellilo, Aber & Muratore (1982) concluded that lignin/N ratios appeared to be the best parameter for predicting litter mass loss rates, and this has been frequently confirmed since. Given the observed changes in litter quality resulting from elevated CO_2 treatment, and the importance of C/N and lignin in controlling litter decomposition rates, it seems inevitable that plants grown under elevated CO_2 will produce litters which will decompose more slowly. When one considers the important role of decomposition in controlling soil C stores, it becomes necessary to evaluate the little work which has been performed on the effects of elevated CO_2 on litter decomposition rates.

O'Neill & Norby (1991) performed a field litter bag experiment using yellow-poplar leaf litters derived from enriched and ambient CO_2 atmospheres, and reported no changes in decay rates for the two litter types. However, in this study, the leaves derived from both elevated and ambient CO_2 treatments did not differ in litter quality parameters (e.g. C/N and lignin/N). In the same year, Coûteaux et al. (1991) studied the decomposition of sweet chestnut leaves derived from ambient and elevated CO_2 environments using laboratory units with food-webs of differing complexities. These workers observed that litter derived from plants grown under elevated CO_2 had decreased decomposition rates, (suggested to be a consequence of changes in litter C/N ratios), but only for litter incubated with microflora and protozoa. The reverse was observed for litter incubated with more complex fauna, with elevated CO_2 enhancing decomposition rates. The limited amounts of leaf litter available for this experiment precluded field studies and, given the conflicting nature of the results, it is difficult to predict field responses. Field studies with the full complement of soil biota are needed to resolve this paradox, and microcosm (Cotrufo et al., 1994) and field litter bag experiments (Cotrufo & Ineson, 1996) using litter from birch and other species have reinforced the results from the simple community studies of Coûteaux et al. (1991). Decreased decomposition rates for root materials derived from elevated CO_2 treatments have also been shown (Cotrufo & Ineson, 1995; Gorissen et al., 1995).

Taylor & Ball (1994) did not observe any significant differences in the biodegradability of C_4 plant tissues derived from either elevated or ambient CO_2 regimes, but analysis of the plant material did not reveal any significant effects of the CO_2 treatment on tissue C/N ratios in their study.

DIRECT EFFECTS OF ELEVATED CO_2 ON SOIL ORGANISMS

The concentration of CO_2 in soil air usually ranges between 0.3 and 3%, but may reach values of 5-10% or higher, generally increasing with soil depth. The CO_2 in soil air exchanges continuously with the CO_2 in the external atmosphere, yet the concentration in soil air is dominated by factors controlling internal production. In a study of CO_2 gradients through a soil-plant system, Schwartz & Bazzaz (1973) measured CO_2 concentrations from above a canopy down to a soil depth of 2 m, and a doubling in the CO_2 level was observed at a soil depth of 10 cm. These authors identified two processes controlling CO_2 levels in the soil air: in the spring, CO_2 was released by upwards diffusion from ground-water rich in dissolved CO_2 whilst, in the summer, biological activity of soil organisms became the major controlling factor. On the basis that CO_2 concentrations in the soil are much higher than CO_2 levels in the immediate atmosphere, and that soil CO_2 concentration is not directly dependent upon levels of CO_2 in the atmosphere, we may conclude that even a doubling in the concentration of atmospheric CO_2 is unlikely to induce any direct significant change in the composition of soil air.

The effect of CO_2 concentrations on microbial activity in the soil has not been extensively investigated but the dogma that soil organisms are relatively insensitive to high CO_2 levels has stood the test of time. MacFadyen (1973) observed that, in light sandy soil, CO_2 concentrations of the soil air as high as 2.5% reduced the metabolic rate of microorganisms, whilst microbial activity was suppressed at CO_2 concentrations of 8%. This inhibitory effect is recognized to be a property of soil type, and to be related to the microflora of the particular habitat. In a detailed review of the effects of soil air composition on fungi, Taback & Cooke (1968) reported that the levels of CO_2 which inhibit or stimulate fungal growth and metabolism vary with species of fungi and are dependent on other environmental factors. However, most of the studies cited in that review used very high CO_2 levels, ranging from 1 to 100%. In a study of the responses of wood-decomposing fungi to high CO_2 levels, Bavendamn (1928) observed that CO_2 concentration in the soil air of about 20% only slightly retarded the growth of wood-decomposing fungi, with cellulose-decomposing fungi being the most sensitive to high CO_2 levels, and lignin-decomposing fungi being relatively insensitive. Soil animals are frequently adapted to living at very high levels of CO_2 (Swift, Heal & Anderson, 1979) and critical CO_2 concentrations for some groups of soil fauna are reported in Table 2. In a review of the effects of elevated CO_2 on insects, Nicolas & Sillans (1989) identified insects which inhabit environments with naturally high CO_2 contents of 2-5 %, such as those found in decomposing dung, soil covered by ice or snow and underneath the bark of trees. Insects living in warm, moist soil are found to be adapted to concentrations of CO_2 up to 3-5% (Ghilarov, 1983), and some can tolerate CO_2 levels above 35% (Vannier, 1983). Among the micro-fauna, enchytraeids are one of the groups reputed to be most affected by high

Table 2. Effects of elevated CO$_2$ concentrations of soil air on soil fauna.

Group	Species	Effect	CO$_2$ (%)	Reference
Microbes	•	inhibited respiration	7 (pH 4)	MacFadyen (1973)
Fungi	*Fusarium acuminatum*	50% mass decrease	10 (pH 6)	Macauley & Griffin (1969)
	Cochliobolus spicifer	63% mass decrease	10 (pH 6)	
	Cochliobolus sativus	increased production	10 (pH 6)	
	Chaetomium sp.	increased production	10 (pH 4-7)	
Acari	*Damaeuso nustus*	50% population	9.1	Moursi (1962)
	Carabodes femoralis	death	7.9	
	Achipteria coleoptrata		7.5	
	Xenillus tegeocranus		6.7	
Collembola	*Onychiurus granulosas*	50% population	1.7	Moursi (1962)
	Hypogastrura bengtssoni	death	6.7	
	Isotomina thermophila		17	
Isopodi	*Pachygaster* sp.	50 % population	30	Moursi (1962)
	Porcellio scaber	death	1.2	

concentrations of CO$_2$ (Kevan, 1968). Because elevated levels of atmospheric CO$_2$ are unlikely to induce any direct changes in the CO$_2$ concentration of the soil air, and such changes would be insignificant with respect to the normal metabolism of soil organisms, enriched atmospheric CO$_2$ is expected to have no *direct* effects on soil C transfers.

EFFECTS ON SOIL C STORES

Soils are estimated to contain 1400 to 1700 Pg of C, two to three times the amount in the living terrestrial biota (Houghton & Skole, 1990), and a diagram of the distribution of C pools between plant biomass and soils for the major terrestrial ecosystems is reported in Fig. 4, re-drawn from Anderson (1992). Accurate predictions of the changes in soil C stores under an atmosphere of elevated CO$_2$ are difficult to make, but must allow for changes in the quality and quantity of plant material entering the soil, both of which may lead to an increase in soil C stores.

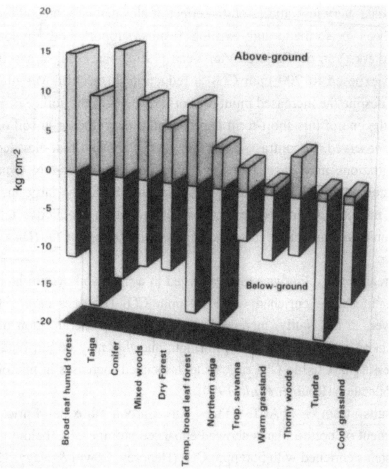

Fig. 4. Carbon pools above and below ground in major ecosystem types. Data from Anderson (1992).

However, in addition to the specific increases in below-ground plant biomass, it is anticipated that there will be increases in root exudation (Van Veen *et al.*, 1991). It is a matter of debate whether such increases in root exudation will exert positive or negative effects on existing soil organic matter (SOM) stores, and two conflicting hypotheses have emerged:

i)   the stimulation of microbial activity caused by increased C input via root exudation will lead to enhanced SOM turnover (the so-called 'priming effect'), resulting in no net change, or even a decrease, in soil C stores;

ii)   elevated C inputs from increased root exudation will not induce an increase in turnover of existing SOM, and will therefore lead to an increase in the organic matter content of soils.

To date, very few studies on the effect of elevated CO_2 on soil C stores have been performed, and existing results support both hypotheses. Lekkerkerk, Van de Geijn & Van Veen (1990) observed, in a soil-plant system exposed to 700 ppm CO_2, a reduction in the turnover of native SOM, despite the increased input of root-derived C to the soil.

At the end of this short-term experiment, a net increase in soil organic C was observed. In contrast, Zak et al. (1993) showed that elevated CO_2 concentrations may have a positive effect on soil C and N dynamics, increasing microbial biomass and N availability. Similarly, large increases in soil microbial C and N were observed for both a productive tall herb community and an acidic grassland exposed to 700 ppm CO_2 (Diaz et al., 1993).

Increased C cycling was also observed in a plant/soil system incubated in an atmosphere enriched with 700 ppm CO_2 (Rouhier et al., 1994). However, in this study, increased organic matter mineralization did not result in an increase in N mineralization, and this may be attributable to changes in plant tissue C/N ratios and associated increases in microbial N immobilization (Rouhier et al., 1994).

Results from a FACE (Free Air Carbon dioxide Enrichment) experiment on cotton plants showed enhanced storage of C below ground for systems enriched with 550 ppm CO_2 (Hendrey, Lewin & Nagy, 1993).

However, increased soil respiration was also observed in experimental rings exposed to elevated CO_2, and this was thought to be the result of increased root and microbial respiration, the latter being stimulated by enhanced release of labile organic compounds. Increased fungal counts in soils exposed to high CO_2 levels support this hypothesis (Hendrey et al., 1993).

EFFECTS ON SOIL PHYSICAL CLIMATE

The effects that CO_2 can exert on soil physical properies via plants are extremely difficult to quantify and predict. Increases in plant productivity may potentially result in lower soil temperature and moisture through interception by a larger canopy, but stomatal responses tend to raise soil moisture contents. However, soils cropped with cotton under ambient and elevated (550 ppm) CO_2 concentrations The effects that CO_2 can exert on soil physical properties *via* plants are extremely at the US FACE did not differ in these properties, with no significant differences in soil water content, pH and bulk density being observed between CO_2 treatments (Rogers, Prior & O'Neil, 1992).

ELEVATED CO_2 AND DECOMPOSITION OF TREE LEAF LITTER DERIVED FROM A CO_2 SPRING

The evidence presented above suggests that vegetation exposed to elevated concentrations of CO_2 will produce a litter which has a higher C/N ratio and, therefore, will decompose more slowly than that produced under current ambient concentrations. This appears to be one of the most consistent effects of elevated CO_2 and, if it is as significant as indicated through controlled experimentation, it should be reflected in materials naturally growing around elevated CO_2 sources.

The Bossoleto (Terme San Giovanni, Siena, Italy) is a natural CO_2 spring, and a full description of the site is provided by Körner & Miglietta (1994). A large area of the Bossoleto basin is forested, with oak trees (*Quercus ilex* and *Quercus pubescens*) being the most common species; litter from these trees was used for the work described below.

Senescent leaves were collected directly from *Quercus pubescens* (Mill.) trees in early January 1993, from two locations: the Bossoleto and Poggio Santa Cecilia, a nearby small deciduous woodland, used as a control. Obviously, one of the main difficulties associated with using natural springs for CO_2 enrichment studies is the location of a suitable reference area and Poggio Santa Cecilia has been used previously as a reference site for other studies centred around the Bossoleto, and was suggested as most appropriate for the current work (A. Raschi, pers. comm.).

Three oak trees were sampled at each site and leaves were collected at the same height, in order to have leaves which had been exposed to similar CO_2 atmospheric concentrations. For the Bossoleto, CO_2

concentrations at the location where leaves were collected ranged between 600 and 700 ppm, which was confirmed at the time of leaf collection using a portable infrared gas analyser (IRGA, PPSystem EGM-1). After collection, leaves were air-dried and stored in plastic bags, prior to being shipped to the UK.

Leaves, with petioles removed, were used for chemical analysis and for the decomposition study. Leaves were kept separate by site and tree sampled, leading to three replicates batches of leaves for each site. A leaf sub-sample for each replicate was milled in a liquid nitrogen mill and analysed for total C and N. Analyses were performed by combustion in an elemental analyser (Carlo Erba 1106). Leaf water contents were measured on oven-dried sub-samples (80°C).

The decomposition study was performed in laboratory units, with samples (1.5 g) of air-dry leaf litter being placed in 150 cm^3 Erlenmeyer flasks, flasks being plugged with cotton wool to permit gas exchange (Dursun *et al.*, 1993); one unit was established for each litter sample, with a total of three units for each site. Litters were inoculated with 3 cm^3 of an L-layer suspension filtered to 1 mm, and re-hydrated by soaking in 100 cm^3 deionized water overnight. Units were then drained, total weight taken and incubated in a constant temperature room at 15°C. Litters were kept moist by addition, every 14 days and before the measurement of CO$_2$ evolution, of deionized water to a pre-determined weight.

Soon after draining, and every two weeks thereafter, for an experimental period of 16 weeks, litter respiration was measured. After replacing plugs with rubber stoppers (Suba-Seal, Barnsley, UK), flasks were flushed with CO$_2$-free air, and incubated for about 3 hours (time accurately noted) at 15°C. CO$_2$ concentration in each flask was then measured on 1 cm^3 gas samples by Infrared Gas Analysis (IRGA, Type 225 Mk3, The ADC Ltd., Hoddesdon, UK). Triplicate samples were taken from each flask. At the end of the experiment, litters were removed from the flasks and mass determined after drying in an oven at 80°C.

The differences between litters derived from the reference and enriched sites for C and N concentrations, C/N ratio, respiration and weight remaining (WTREM) were analysed by Student-*t* test. The N content of oak leaves derived from the CO_2 spring and the reference site are reported in Table 3. For both litter types, N concentration values were within normal ranges (Rodin & Basilevich, 1967), and no signs of N deficiency were observed. Whilst C content was the same in both litters, the Bossoleto leaf litter had a lower concentration of N, and concomitant higher C/N ratios (Table 3). However, these observed trends in litter quality were not statistically significant.

TABLE 3. Nitrogen composition of oak leaves derived from the Bossoleto (enriched) and the reference site (control). Values are means ($n=3$), with standard errors in parentheses. Differences were analysed by Student - t, and significance levels are reported (n.s. = not significant).

Site	N (%)	C (%)	C/N
Enriched	0.97 (0.26)	49 (0.88)	55 (9.95)
Reference	1.20 (0.22)	49 (0.67)	45 (9.89)
Sources of variance			
Site	n.s.	n.s.	n.s.

Respiration rates of litters derived from the natural enriched CO2 and from the control sites are reported in Fig. 5. Litter derived from the spring showed lower respiration rates throughout the experiment, and on two occasions the differences were statistically significant: at the 6th and the 8th weeks. Both litters showed high respiration rates at the beginning of the incubation, which remained almost constant from the 4th to the 14th week, and reduced at the end of the experiment.

At the end of the 16 weeks of laboratory incubation there appeared to be greater mass loss for the control litters but differences in the weight remaining between litters derived from the enriched and control sites were not statistically significant (Fig. 6).

In the present study, long-term exposure to naturally enriched CO_2 atmosphere appeared to have led to changes in litter quality showing similar trends to those found from controlled fumigation experiments, with lower N contents and enhanced C/N ratios. Such observations are encouraging, in that they support data derived from short-term fumigations.

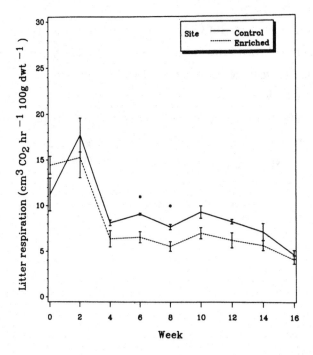

FIG. 5. Litter respiration versus time for oak litters derived from a natural CO_2 enriched site (enriched), and from a reference site (control). Vertical bars indicate standard deviation. Significant differences are given after Student's t test, *$p<0.05$.

The results obtained from this study suggest that leaf litters generated from mature trees, grown for their entire life under elevated CO_2 regimes, may decompose more slowly than leaf litters from similar trees grown at ambient CO_2 levels.

The difference in litter decay rates appeared to be linked to changes in litter quality (e.g. N content and C/N ratio) and, again, such results are in agreement with the observations reported above.

However, it must be emphasised that results from such field surveys around vents may be due to factors other than the known differences in CO_2 concentrations.

The data presented here suggest that further investigations of C turnover rates at these natural CO_2 vents are warranted. Not only are natural vents a useful long-term resource for CO_2 enrichment studies, but they are of intrinsic scientific interest as extreme environments. They offer the opportunity of providing inexpensive field incubation sites for transplanted or imported vegetation, as well as potential sources of CO_2 for controlled field fumigation experiments.

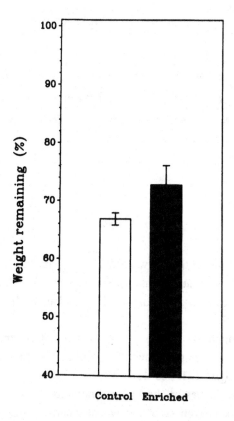

FIG. 6. Litter weight remaining at the end of the experiment for oak litter derived from a natural CO_2 enriched site (enriched), and from a reference site (control). Vertical bars indicate standard errors.

The present work has shown that elevated concentrations of CO_2 affect litter decomposition through changes in litter quality. In particular, elevated CO_2 decreases litter N concentrations with a resulting increase in C/N and lignin/N ratios, leading to a slow down in litter decomposition rates. Such observations are true for both leaf and root litters, and for laboratory as well as field incubations. CO_2 treatment does not appear to exert any effect on litter decay rates when there are no measurable changes in leaf chemical composition. Furthermore, it has been identified that elevated CO_2 is unlikely to exert any direct effects on soil biological activity, and that only effects mediated *via* changes in plants are likely to occur. The overall conclusion must be that soil C stores are increasing, and will continue to do so, under higher CO_2 concentrations.

ACKNOWLEDGEMENTS

We thank Dr. A. Raschi and his collaborators at the CNR-IATA, Florence for provision of the oak leaves used in the current work. The study was funded under the UK Natural Environmental Research Council TIGER programme.

REFERENCES

Ågren, G.I. (1994) The interaction between CO_2 and plant nutrition: comments on a paper by Coleman, McConnaughay and Bazzaz. *Oecologia,* **98**, 239-240.

Anderson, J.M. (1992) Responses of soils to climate change. *Advances in Ecological Research,* **22**, 163-211.

Barnes, J.D. & Pfirrmann, T. (1992) The influence of CO_2 and O_3, singly and in combination, on gas exchange, growth and nutrient status of radish (*Raphanus sativus*). *New Phytologist,* **121**, 403-412.

Bavendamn, W. (1928) Studies on the living conditions of wood-destroying fungi. A contribution to the sensitiveness of our woody plants to disease. *Zentralbl Bakt,* **II.75**, 426-452, 503-533.

Bazzaz, F.A. & Miao, S.L. (1993) Successional status, seed size and responses of tree seedlings to CO_2, light and nutrients. *Ecology,* **74**, 104-112.

Berg, B. (1984) Decomposition of root litter and some factors regulating the process: long-term root litter decomposition in a Scots pine forest. *Soil Biology & Biochemistry,* **16**, 609-617.

Berryman, C.A., Eamus, D. & Duff, G.A. (1993) The influence of CO_2 enrichment on growth, nutrient content and biomass allocation of *Maranthes corymbosa*. *Australian Journal of Botany,* **41**, 195-209.

Bhattacharya, N.C., Biwas, P.K., Bhattacharya, S., Sionit, N. & Strain, B.R. (1985) Growth and yield response of sweet potato to atmospheric CO_2 environment. *Crop Science,* **25**, 975-981.

Brown, K.R. (1991) Carbon dioxide enrichment accelerates the decline in nutrient status and relative growth rate of *Populus tremuloides* Michx. seedlings. *Tree Physiology,* **8**, 161-173.

Chu, C.C., Coleman, J.S. & Mooney, H.A. (1992) Control of biomass partitioning between roots and shoots: Atmospheric CO_2 enrichment and the acquisition and allocation of carbon and nitrogen in wild radish. *Oecologia,* **89**, 580-587.

Cipollini, M.L., Drake, B.G. & Whigham, D. (1993) Effects of elevated CO_2 on growth and carbon/nutrient balance in the deciduous woody shrub *Lindera benzoin* (L.) Blume (Lauracea). *Oecologia*, **96**, 339-346.

Coleman, J.S., McConnaughay, K.D. M. & Bazzaz, F.A. (1993) Elevated CO_2 and plant nitrogen use: is reduced tissue nitrogen concentration size-dependent? *Oecologia*, **93**, 195-200.

Conroy, J. & Hocking, P. (1993) Nitrogen nutrition of C_3 plants at elevated atmospheric CO_2 concentrations. *Physiologia Plantarum*, **89**, 570-576.

Cotrufo, M.F., Ineson, P. & Rowland, A.P. (1994) Decomposition of tree leaf litters grown under elevated CO_2 : Effect of litter quality. *Plant & Soil*, **163**, 121-130

Cotrufo, M.F. & Ineson, P. (1995) Effects of enhanced atmospheric CO_2 and nutrient supply on the quality and subsequent decomposition of fine roots of *Betula pendula* Roth. and *Picea sitchensis* (Bong.) Carr. *Plant & Soil*, **170**, 267-277.

Cotrufo, M.F. & Ineson, P. (1996) Elevated CO_2 reduces field decomposition rates of *Betula pendula* Roth. leaf litter. *Oecologia*, in press.

Coûteaux, M.M., Mousseau, M., Célérier, M.L. & Bottner, P. (1991) Increased atmospheric CO_2 and litter quality: decomposition of sweet chestnut leaf litter with animal food webs of different complexities. *Oikos*, **61**, 54-64.

Curtis, P.S., Drake, B.G., Leadley, P.W., Arp, W.J. & Whigham, D.F. (1989a) Growth and senescence in plant communities exposed to elevated CO_2 concentrations on an estuarine marsh. *Oecologia*, **78**, 20-26.

Curtis, P.S., Drake, B.G. & Whigham, D.F. (1989b) Nitrogen and carbon dynamics in C_3 and C_4 estuarine marsh plants grown under elevated CO_2 *in situ*. *Oecologia*, **78**, 297-301.

Diaz, S., Grime, J.P., Harris, J. & McPherson, E. (1993) Evidence of a feedback mechanism limiting plant response to elevated carbon dioxide. *Nature*, **364**, 616-617.

Dursun, S., Ineson, P., Frankland, J.C. & Boddy, L. (1993) Sulphite and pH effects on CO_2 evolution from decomposing angiospermous and coniferous tree leaf litters. *Soil Biology & Biochemistry*, **25**, 1513-1525.

Eamus, D. & Jarvis, P.G. (1989) The direct effects of increase in the global atmospheric CO_2 concentration on natural and commercial temperate trees and forests. *Advances in Ecological Research*, **19**, 1-55.

El-Kohen, E.I., Rouhier, H. & Mousseau, M. (1992) Changes in dry weight and nitrogen partitioning induced by elevated CO_2 depend on soil nutrient availability in sweet chestnut (*Castanea sativa* Mill.). *Annales des Science Foresterierés*, **49**, 83-90.

Flanagan, P.W. & Van Cleve, K. (1983) Nutrient cycling in relation to decomposition and organic-matter quality in taiga ecosystems. *Canadian Journal of Forest Research*, **13**, 795-817.

Ghilarov, M.S. (1983) Some essential physiological characters of soil invertebrates. In *New Trends in Soil Biology* (eds. P. Lebrun, H.M. André, A. De Medts, C. Gregoire-Wibo & G. Wauthy), pp. 315-320. Dieu-Brichart, Ottiguies-Louvain-la-Neuve.

Gifford, R.M. (1994) The global carbon cycle: a viewpoint on the missing sink. *Australian Journal of Plant Physiology*, **21**, 1-15.

Gorissen, A., Van Ginkel, J.H., Keurentjes, J.J.B. & Van Veen, J.A. (1995) Grass root decomposition is retarded when grass has been grown under elevated CO_2. *Soil Biology & Biochemistry*, **27**, 117-120.

Hendrey, G.R., Lewin, K.F. & Nagy, J. (1993) Free air carbon dioxide enrichment: development, progress, results. *Vegetatio*, **104/105**, 17-31.

Houghton, R.A. (1993) Is carbon accumulating in the Northern temperate zone? *Global Biogeochemical Cycles*, **7**, 611-617.

Houghton, R.A. & Skole, D.L. (1990) Carbon. In *The Earth as Transformed by Human Action* (eds. B.L. Tarner II, W.C. Clark, R.W. Kates, J.F. Richards, J.T. Mathews & W.B. Meyer), pp. 393-408. Cambridge University Press, Cambridge.

Johnson, R.H. & Lincoln, D.E. (1991) Sagebrush carbon allocation patterns and grasshopper nutrition: the influence of CO_2 enrichment and soil mineral limitation. *Oecologia*, **87**, 127-134.

Julkunen-Tiitto, R., Tahvanainen, J. & Silvola, J. (1993) Increased CO_2 and nutrient status changes affect phytomass and production of plant defensive secondary chemicals in *Salix myrsinifolia* (Salisb.). *Oecologia*, **95**, 495-498.

Kevan, K. McE. (1968) *Soil Animals*. HF & G Witherby, London.

Körner, C. & Miglietta, F. (1994) Long term effects of naturally enriched CO_2 on mediterranean grassland and forest trees. *Oecologia*, **99**, 343-351.

Lambers, H. (1993) Rising CO_2, secondary plant metabolism, plant-herbivore interactions and litter decomposition. *Vegetatio*, **104/105**, 263-271.

Lekkerkerk, L.J.A., Van de Geijn, S.C. & Van Veen, J.A. (1990) Effect of elevated atmospheric CO_2-levels on the carbon economy of a soil planted with wheat. In *Soils and the Greenhouse Effect* (ed. A.F. Bouwmann), pp. 423-429. John Wiley & Sons, Chichester.

Lindroth, R.L., Kinney, K.K. & Platz, C.L. (1993) Responses of deciduous trees to elevated atmospheric CO_2: productivity, and insect performance. *Ecology*, **74**, 763-777.

Luxmoore, R.J., Wullschleger, S.D. & Hanson, P.J. (1993) Forest responses to CO_2 enrichment and climate warming. *Water Air & Soil Pollution*, **70**, 309-323.

Macauley, B.J. & Griffin, D.M. (1969) Effect of carbon dioxide and the bicarbonate ion on the growth of soil fungi. *Transactions British Mycological Society*, **53**, 223-228.

MacFadyen, A. (1973) Inhibitory effects of carbon dioxide on microbial activity in soil. *Pedobiologia*, **13**, 140-149.

Mason, C.F. (1977) *Decomposition*. The Camelot Press Ltd., Southampton.

Melillo, J.M. (1983) Will increases in atmospheric CO_2 concentrations affect decay process? In *The Ecosystem Center. 1983 Annual Report*, pp. 10-11. Marine Biological Laboratory, Woods Hole, Massachusetts, USA.

Melillo, J.M., Aber, J.D. & Muratore, J.F. (1982) Nitrogen and lignin control of hardwood leaf litter decomposition dynamics. *Ecology*, **63**, 621-626.

Mooney, H.A., Drake, B.G., Luxmoore, R.J., Oechel, W.C. & Pitelka, L.F. (1991) Predicting ecosystem responses to elevated CO_2 concentrations. *Bioscience*, **41**, 96-104.

Moursi, A.A. (1962) The lethal doses of CO_2, N_2, NH_3 and H_2S for soil Arthropoda. *Pedobiologia*, **2**, 9-14.

Nicolas, G. & Sillans, D. (1989) Immediate and latent effects of carbon dioxide on insects. *Annual Review of Entomology*, **34**, 97-116.

Norby, R.J., Pastor, J. & Melillo, J.M. (1986a) Carbon-nitrogen interactions in CO_2-enriched white oak: physiological and long-term perspectives. *Tree Physiology*, **2**, 233-241.

Norby, R.J., O'Neill, E.G. & Luxmoore, R.J. (1986b) Effects of atmospheric CO_2 enrichment on the growth and mineral nutrition of *Quercus alba* seedlings. *Plant Physiology*, **82**, 83-89.

Oberbauer, S.F., Sionit, N., Hastings, S.J. & Oechel, W.C. (1986) Effects of CO_2 enrichment and nutrition on growth, photosynthesis, and nutrient concentration of Alaskan Tundra plant species. *Canadian Journal of Botany*, **64**, 2993-2998.

O'Neill, E.G., Luxmoore, R.J. & Norby, R.J. (1987) Elevated atmospheric CO_2 effects on seedling growth, nutrient uptake, and rhizosphere bacterial population of *Lirodendron tulipifera* L. *Plant & Soil*, **104**, 3-11.

O'Neill, E.G. & Norby, R. J. (1991) First year decomposition of Yellow-poplar leaves produced under CO_2 enrichment. *Bulletin of the Ecological Society of America*, **72**, 208.

O'Neill, E.G. (1994) Responses of soil biota to elevated atmospheric carbon dioxide. *Plant & Soil*, **165**, 55-65.

Overdieck, D. & Reining, E. (1986) Effect of atmospheric CO_2 enrichment on perennial ryegrass (*Lolium perenne* L.) and white clover (*Trifolium repens* L.) competing in managed model-ecosystems. II. Nutrient uptake. *Acta Oecologica, Oecologia Plantarum*, **7**, 367-378.

Overdieck, D., Reid, C. & Strain, B.R. (1988) The effects of preindustrial and future CO_2 concentrations on growth, dry matter production and the C/N rationship in plants at low nutrient supply: *Vigna unguiculata* (cowpea), *Abelmoschus esculentus* (okra) and *Raphanus sativus* (radish). *Vereinigung für Angewandte Botanik*, **62**, 119-134.

Owensby, C.E., Coyne, P.I. & Aven, L.M. (1993) Nitrogen and phosphorus dynamics of a tallgrass prairie ecosystem exposed to elevated carbon dioxide. *Plant, Cell & Environment*, **16**, 843-850.

Peñuelas, J. & Matamala, R. (1990) Change in N and S leaf content, stomatal density and specific leaf area of 14 plant species during the last three centuries of CO_2 increase. *Journal of Experimental Botany*, **41**, 1119-1124.

Rodin, L.E. & Bazilevich, N.I. (1967) *Production and Mineral Cycling in Terrestrial Vegetation.* Oliver & Boyd, London.

Rogers, H.H., Prior, S.A. & O'Neill, E.G. (1992) Cotton root and rhizosphere responses to Free-Air CO_2 Enrichment. *Critical Reviews in Plant Sciences*, **11**, 251-263.

Rouhier, H., Billès, G., El Kohen, A., Mousseau, M. & Bottner, P. (1994) Effects of elevated CO_2 on carbon and nitrogen distribution within a tree (*Castanea sativa* Mill.)-soil system. *Plant & Soil*, **162**, 281-292.

St. Omer, L. & Horvath, S.M. (1983) Elevated carbon dioxide concentrations and whole plant senescence. *Ecology*, **64**, 1311-1314.

Santantonio, D. & Grace J.C. (1987) Estimating fine root production and turnover from biomass and decomposition data: a compartment-flow model. *Canadian Journal of Forest Research*, **17**, 900-908.

Schwartz, D.M. & Bazzaz, F.A. (1973) In *situ* measurements of carbon dioxide gradients in a plant-soil system. *Oecologia*, **12**, 161-167.

Swift, M.J., Heal, O.W. & Anderson, J.M. (1979) *Decomposition in Terrestrial Ecosystems.* Blackwell Scientific Publications, Oxford.

Taback, H.H. & Cooke, W.B. (1968) The effects of gaseous environments on the growth and metabolism of fungi. *Botanical Review*, **34**, 126-252.

Tans, P.P., Fung, I.Y. & Takahashi, T. (1990) Observational constraints on the global atmospheric CO_2 budget. *Science*, **247**, 1431-1438.

Taylor, J. & Ball, A.S. (1994) The effect of plant material grown under elevated CO_2 on soil respiratory activity. *Plant & Soil*, **162**, 315-318.

Tissue, D.T. & Oechel, W.C. (1987) Response of *Eriophorum vaginatum* to elevated CO_2 and temperature in the Alaskan tussock tundra. *Ecology*, **68**, 401-410.

Vannier, G. (1983) The importance of ecophysiology for both biotic and abiotic studies of the soil. In *New Trends in Soil Biology* (eds. P. Lebrun, H.M. André, A. De Medts, C. Gregoire-Wibo & G. Wauthy), pp. 289-314. Dieu-Brichart, Ottiguies-Louvain-la-Neuve.

Van Veen, J.A., Liljeroth, E., Lekkerkerk, L.J.A. & Van de Geijn, S.C. (1991) Carbon fluxes in plant-soil systems at elevated atmospheric CO_2 levels. *Ecological Applications*, **1**, 175-181.

Vu, J.C.V., Allen, L.H. & Bowes, J.R.G. (1989) Leaf ultrastructure, carbohydrates and protein of soybeans grown under CO_2 enrichment. *Environmental & Experimental Botany*, **29**, 141-147.

Wong, S.C. (1979) Elevated atmospheric partial pressure of CO_2 and plant growth. I. Interactions of nitrogen nutrition and photosynthetic capacity in C_3 and C_4 plants. *Oecologia*, **44**, 68-74.

Woodin, S., Graham, B., Killick, A., Skiba, U. & Cresser, M. (1992) Nutrient limitation of the long term response of heather [*Calluna vulgaris* (L.) Mull.] to CO_2 enrichment. *New Phytologist*, **122**, 635-642.

World Resources Institute (1992) *World Resources 1992-93*. Oxford University Press, New York.

Wullschleger, S.D., Post, W.M. & King, A.W. (1995) On the potential for a CO_2 fertilization effect in forest trees. Estimates of a biotic growth factor based on 58 controlled-exposure studies. In *Biospheric Feedbacks in the Global Climatic System: Will Warming Feed the Warming?* (eds. G.M. Woodwell & T. Mackensie), Oxford University Press, New York pp. 85-107.

Zak, D.R., Pregitzer, K.S., Curtis, P.S., Teeri, J.A., Fogel, R. & Randlett, D.L. (1993) Elevated atmospheric CO_2 and feedback between carbon and nitrogen cycles. *Plant & Soil*, **151**, 105-117.

Index

Note: **References in bold** indicate major chapter headings.